Lecture Notes in Mathematics 1490

Klaus Metsch

Linear Spaces with Few Lines

Springer-Verlag
Berlin Heidelberg New York
London Paris Tokyo
Hong Kong Barcelona
Budapest

Author

Klaus Metsch
Mathematisches Institut
Justus-Liebig-Universität
Arndtstraße 2, W-6300 Gießen, FRG

Mathematics Subject Classification (1991): 51E20

ISBN 3-540-54720-7 Springer-Verlag Berlin Heidelberg New York
ISBN 0-387-54720-7 Springer-Verlag New York Berlin Heidelberg

Typesetting: Camera ready by author
Printing and binding: Druckhaus Beltz, Hemsbach/Bergstr.
46/3140-543210 - Printed on acid-free paper

FÜR GERTRUD

Franz kam sein Leben zwischen den Büchern unwirklich vor. Er sehnte sich nach dem wirklichen Leben, nach Berührung mit anderen Menschen, die an seiner Seite gingen, er sehnte sich nach ihrem Geschrei. Es war ihm nicht klar. daß gerade das, was ihm unwirklich schien (die Arbeit in der Einsamkeit von Studierzimmer und Bibliothek), sein wirkliches Leben war, während die Umzüge, die für ihn die Wirklichkeit darstellten, nur Theater waren, ein Tanz, ein Fest, mit anderen Worten ein Traum.

(Milan Kundera: Die unerträgliche Leichtigkeit des Seins)

INTRODUCTION

It is known since 40 years that a linear space has at least as many lines as points with equality only if it is a generalized projective plane. This result of de Bruijn and Erdős (1948) led to the conjecture that every linear space with "few lines" can be obtained from a certain projective plane by changing only a small part of its structure. It is surprising that it took more than 20 years until Bridges (1972) showed that every linear space with $b = v+1 \neq 6$ (b the number of lines, v the number of points) is a punctured projective plane. However, since then many results have been obtained. It is the main purpose of this paper to study systematically this embedding problem. In particular, we shall collect the old results and present quite a few new ones. We shall, however, also study linear spaces with few lines which have no natural embedding in a projective plane.

When studying finite linear spaces which have a chance to be embeddable in a projective plane of order n, it is sensible to suppose that $b \leq n^2+n+1$ (which is the number of lines in a projective plane of order n) and $v \geq (n-1)^2+(n-1)+2 = n^2-n+2$ (this is due to the fact that a projective plane of order $n-1$ has n^2-n+1 points and such a plane is certainly not embeddable in a projective plane of order n).

In the first chapter, we shall give the definitions and most of the notions needed later. Also the most important examples of linear spaces which we shall deal with are given. At the end of this chapter, the reader will find some basic properties of linear spaces and $(n+1,1)$-designs.

Chapter 2 begins with a new proof of the theorem of de Bruijn and Erdős. This proof uses an easy algebraic method, which will be also useful in the proof of Totten's theorem in chapter 8. The inequality $b \geq v$ has been improved by Erdős, Mullin, Sós, and Stinson (1985) for non-degenerate linear spaces as follows. If n is the unique integer satisfying $n^2-n+1 = (n-1)^2+(n-1)+1 < v \leq n^2+n+1$, then we have

$$b \geq B(v) := \begin{cases} n^2+n-1, & \text{if } v = n^2-n+2 \neq 4 \\ n^2+n, & \text{if } n^2-n+3 \leq v \leq n^2+1 \text{ or } v = 4 \\ n^2+n+1, & \text{if } n^2+2 \leq v. \end{cases}$$

We conclude chapter 2 with a new proof of this result.

In chapter 3, basic properties and results of (n+1,1)-designs can be found. We shall prove a theorem of Vanstone which says that any (n+1,1)-design with $4 \leq n^2 \leq v$ and $b = n^2+n+1$ can be embedded in a projective plane of order n, and we shall also show that $n^2-n+2 \leq v \leq b \leq n^2+n$ implies embeddability.

Now we are ready to study systematically non-degenerate linear spaces L with $v \geq n^2-n+2$ points and $b \leq n^2+n+1$ lines. In chapter 4 we consider the case that some point lies on at most n lines. We shall determine L except in the case where $v = n^2-n+2$ and $b = n^2+n+1$. It turns out that L is one of a few exceptional linear spaces or that L can be extended to a projective plane of order n. This result implies that every point of a non-degenerate linear space with n^2+n+1 lines and at least n^2-n+3 points lies on at least n+1 lines, a consequence which will be very helpful in the next chapters.

The results obtained up to here will now be used to determine all linear spaces with the minimal possible number B(v) of lines. If n is the integer with $n^2-n+2 \leq v \leq n^2+n+1$, then every non-degenerate linear L space with v points and B(v) lines can be embedded in a projective plane of order n unless L is one exceptional linear space with 8 points. This will be shown in chapter 5. We shall also study linear spaces with n^2-n+2 points and n^2+n lines and obtain a classification of all linear spaces with $n^2-n+2 \leq v < b \leq n^2+n$. Only two such linear spaces can not be embedded in a projective plane of order n and both have 8 points.

In chapter 5 we will also determine all linear spaces with n^2-n+1 points and n^2+n lines for which every point has degree at least n+1 and some point has degree at least n+2, $n \neq 4,9$. It turns out that these spaces are related to a complement of a Baer-subplane in an affine plane.

In the following two chapters we consider linear spaces with n^2+n+1 lines and $v \geq n^2-n+2$ points in which every point lies on at least n+1 lines.

We start with the very difficult case that every point lies on exactly n+1 lines. In chapter 3 we have already proved that $v \geq n^2$ implies embeddability. In the last 15 years this bound has been improved many times, for example by Bose and Shrikhande (1973), McCarthy and Vanstone (1977), and Dow (1982, 1983). We shall improve all these bounds again by showing that as well $v \geq n^2-\frac{1}{2}n+6$ as $v \geq n^2 - \frac{1}{2}(\sqrt{5} - 1)n + 17\sqrt{(n/5)}$ implies embeddability.

In chapter 7, we assume that some point lies on more than n+1 lines. Because such linear spaces can not be embedded in a projective plane of order n, it has been conjectured that they do not exist. However, we shall construct an infinite class of counterexamples. Such a counterexample has at most $n^2+1-\sqrt{n}$ points with equality if it is the closed complement of a Baer-subplane in a projective plane of order n. In chapter 7 we shall show that these are essentially the only examples for $v > n^2+1-\frac{1}{2}n$. Since they can not be embedded in a projective plane of order n, we obtain the optimal bound for the embedding of linear spaces in projective planes of order n whenever n is the order of a projective plane having a Baer-subplane.

Chapter 8 starts with a proof of Totten's classification of restricted linear spaces (1976) (linear spaces satisfying $(b-v)^2 \leq v$). We shall improve this result slightly by determining all linear spaces satisfying $(b-v)^2 \leq b$. As corollaries we obtain the classifications of linear spaces with $b = v+1$ of Bridges (1972), $b = v+2$ of de Witte (1976), and $b = v+3$ of Totten (1976). A consequence of the Theorem of Totten is that every linear space satisfying $v \leq n^2+n+1 < b \leq v+n$ is an inflated affine plane of order n, that is an affine plane A of order n together with a linear space with at most n+1 points which is imposed on some of the infinite points of A. It seems likely that this results remains true if one weakens the upper bound $b \leq v+n$ for the number of lines. In the case that the number of points is n^2+n+1, Blokhuis, Schmidt, and Wilbrink (1988) showed that the condition $b \leq n^2+3n+1-4\sqrt{n}$ is strong enough. We shall sligthly improve this result in chapter 9.

In chapter 10, we introduce a class of structures which we call L(n,d). By definition an L(n,d), $1 \leq d \leq n-1$ is an (n+1,1)-design with n^2-d points in which every point lies on n lines of degree n and on a unique line of degree n−d. We shall see that this implies that there is an integer z such that $z(n-d) = d(d-1)$ and $b = n^2+n+z$. It is easy to see that an L(n,d) is a punctured affine plane if $z = 0$ and the complement of a Baer-subplane in a projective plane if $z = 1$. For $z \geq 2$ no examples are known and we give non-existence criteria. However even for $z \geq 2$ these (hypothetical) linear spaces are interesting from many points of view. For example, if $z \geq 2$, then we obtain the following characterization of an L(n,d) with $z(n-d) = d(d-1)$: Suppose L is a linear space with maximal point degree n+1 and $b = n^2+n+z$. Then $v \leq n^2-d$ and equality implies that L is an L(n,d).

Each L(n,d) gives rise to a *closed* L(n,d), which is a linear space with n^2+n+z lines and n^2-d+1 points, and to a *reduced* L(n,d), which is a linear space with n^2-n+2 points, $b = n^2+n+z-1$ lines, and a point of degree n. For $z = 2$, these structures will play a crucial role in the next two chapters. Suppose that n is a positive integer and let d be the positive number defined by $2n = d(d-1)$. In chapter 11, we prove that every linear space with n^2+n+2 lines has at most n^2-d+1 points with equality if and only if it is a closed L(n,d), which implies that d is an integer. In the next chapter we shall show that every linear space with n^2-n+2 points and n^2+n+1 lines which has a point of degree n is a reduced L(n,d) (which implies again that d is an integer) or one of a few exceptional linear spaces. This result solves the unsettled case of chapter 4.

In chapter 13, we consider two particular cases. First, we shall prove that every linear space with $13 < v \leq b \leq 21$ is a near-pencil or can be embedded in the projective plane of order 5. Then we show that every linear space satisfying $31 < v \leq b \leq 43$ is a near-pencil. This includes a new proof for the non-existence of a projective plane of order 6, and the pseudo-complement of an oval in a projective plane of order 6. Our proof does not use any graph theoretical results as did the original proof of de Witte (1977).

An interesting property of linear spaces was discovered by Hanani (1954/55): Every line of maximal degree of a linear space with v points meets at least v-1 other lines. In chapter 15 we prove the following generalisation. Suppose that c is a non-negative integer. Then all but a finite number of linear spaces in which every line of maximal degree meets at most v-1+c other lines can be obtained from a projective plane by removing at most c points or from an affine pane by removing at most c−1 points. The proof applies a very general embedding result, which we give in chapter 14.

Another interesting property of linear spaces was conjectured by Dowling and Wilson (see Erdős, Fowler, Sós and Wilson (1985)): If (p,L) is a non-incident point-line pair of a linear space and if t is the number of lines through p which miss L, then $b \geq v+t$. This generalization of the de Bruijn-Erdős Theorem was proved in (Metsch, 1991c) and we shall present this proof in chapter 16.

In the last chapter we will be concerned with the uniqueness of embeddings. It is conceivable that a linear space may be embedded in two different ways in the same projective plane, or even that it can be embedded in two non-isomorphic planes. We shall, moreover, show that all embeddings considered in this paper are uniquely determined.

The first part of this paper was written while I was visiting the university of Florence, Italy. I would like to thank Prof. A. Barlotti very much for his kind hospitality.

I am very grateful to Prof. A. Beutelspacher, my thesis supervisor, for his help and support in every situation, and I would also like to thank him very much.

C O N T E N T S

1. Definition and basic properties of linear spaces

In this section, we shall give the definitions of the geometrical structures used in this paper. We shall also give examples and basic properties of linear spaces. The most general structure we shall use is the incidence structure.

An *incidence structure* is a triple $I = (p,L,I)$ of a set p of *points*, a set L of *lines*, and a set I of *incidences* satisfying

$$p \cap L = \emptyset \qquad \text{and} \qquad I \subseteq p \times L.$$

If $(p,L) \in I$, then we say that the point p lies on the line L or that L passes through p. If $(p,L) \notin I$, then we say p does not lie on L or p is not a point of L. Further geometric expressions explaining themselves are used.

We only consider incidence structures with a finite number of points and lines. The number of points is denoted by v and the number of lines by b. The number k_L of points on a line L is called the *degree* of L and the number r_p of lines passing through a point p is called the *degree* of p. A *k-line* is a line of degree k. The *parameters* of I are the number of points and lines, the degrees $r_1,\dots,r_v$ of the points, and the degrees $k_1,\dots,k_b$ of the lines.

Two lines L and H are called *parallel* if $L \neq H$ and if L and H do not intersect. A *parallel class* is a set Π of lines such that every point lies on exactly one line of Π.

The incidence structure $I_d = (L,p,\{(L,p) \mid (p,L) \in I\})$ is called *dual* to I.

An incidence structure $I' = (p',L',I')$ is said to be *embedded* in I, if $p' \subseteq p$, $L' \subseteq L$, and $I' \subseteq I \cap (p' \times L')$. An *isomorphism* from I onto an incidence structure $I'' = (p'',L'',I'')$ is a bijection α from $p \cup L$ onto $p'' \cup L''$ mapping points onto points and lines onto lines such that $(p,L) \in I$ if and only if $(\alpha(p),\alpha(L)) \in I''$.

If M is a set of mutually parallel lines of $I = (p,L,I)$ and if ∞ is a new symbol, then $I \infty M$ denotes the incidence structure $(p \cup \{\infty\}, L, I \cup \{(\infty,L) \mid L \in M\})$. We say that $I \infty M$ is the structure obtained from I, if we let the lines of M intersect in an infinite point ∞.

A $v \times b$ matrix $C = (c_{ij})$ is called an *incidence matrix* of I if there are orders $p_1,\dots,p_v$ and $L_1,\dots,L_b$ of the points and lines such that $c_{ij} = 1$ if p_i is a point of L_j and $c_{ij} = 0$ if not.

An incidence structure is called a *partial linear space* if any two distinct points lie on at most one common line, every line has at least two points, and there are at least two lines. A *linear space* is a partial linear space in which for any two distinct points p and q there is a line pq through p and q. A linear space which has a line passing through all but one of its points is called a *near-pencil*, a *degenerate* linear space, or a *degenerate projective plane*. A linear space which is not a near-pencil is called *non-degenerate*. A linear space with $(b-v)^2 \leq v$ is called *restricted*, and a *weakly restricted* linear space is a linear space satisfying $(b-v)^2 \leq b$.

Since a line of a partial linear space $L = (p,L,I)$ is uniquely determined by its points, we shall identify a line with the set of its points. We write $p \in L$ instead of $(p,L) \in I$, and $L = (p,L)$ instead of $L = (p,L,I)$ where L is now seen as a set of subsets of p. The point of intersection of two intersecting lines L and H is denoted by $L \cap H$.

Let p' be a subset of p containing three non-collinear points. Then we can define the linear space $L' = (p', \{L \cap p' \mid L \in L, |L \cap p'| \geq 2\})$ which is *induced by* L *on* p'. If $C := p - p'$, then L' is called *the complement of* C *in* L and it is denoted by L−C. Obviously, L−C is embedded in L. If L is a line of L with $|L \cap p'| \geq 2$, then $L' := L \cap p'$ is a line of L'. In general L and L' do not coincide as sets of points. However, if no confusion is expected, then we identify L and L' as lines. E.g. if $p \in L - L'$, then we call p a point of L' outside of L'.

(Partial) linear spaces with constant point degree n+1 occur very often. It is sometimes more comfortable to consider a little more general structure, the (partial) (n+1,1)-designs. An incidence structure I with constant point degree n+1 is called a *partial (n+1,1)-design*, if any two distinct points are contained in at most one line, and an *(n+1,1)-design*, if any two distinct points p and q lie on a unique line pq. Notice that a line is not uniquely determined by the set of its points. There may be two or more lines of degree 1 which contain the same point p, and also lines of degree 0 are allowed. Even though we regard the set of lines as a family of subsets of the set of points and we write $p \in L$ instead of $(p,L) \in I$.

A *projective plane* is a linear space P for which there is an integer $n \geq 2$ such that every point and line has degree n+1. The integer n is called the *order* of P. It is easy to see that P has n^2+n+1 points and lines and that any two lines of P intersect. A *generalized projective plane* is a near-pencil or a projective plane. An *affine plane of order* n is the complement of a line L in a projective plane of order n. The n+1 points of L are called the *infinite points* of the affine plane. It is known that a projective plane of order n exists whenever n is the power of a prime.

Suppose that $n = m^2$ is a perfect square and that P is a projective plane of order n with a *Baer-subplane* $B = (q,\Pi)$, i.e, B is a projective plane of order m which is embedded in P. Then $L := P-q$ is also called the *complement of the Baer-subplane* B *in the projective plane* P. The lines of Π, considered as lines of L, form a parallel class of L. $L \infty \Pi$ is called the *closed complement of* B *in* P. It has n^2-m+1 points and n^2+n+1 lines. Furthermore, the point ∞ has degree $|\Pi| = n+\sqrt{n}+1$.

Now suppose that $n = m^2$ is a perfect square and that A is an affine plane of order n with a *Baer-subplane* $B' = (q',\Pi')$, i.e. B' is an affine plane of order m which is embedded in A. Then $L' := A-q'$ is also called the *complement of the Baer-subplane* B' *in the affine plane* A. Suppose that the lines of Π', considered as lines of L', form a parallel class of L'. Then $L' \infty \Pi$ is called the *closed complement of* B' *in* A. It has n^2-n points and n^2+n lines. Furthermore, the point ∞ has degree $|\Pi'| = n+\sqrt{n}$.

If C is the set consisting of the 2n+1 points of two lines of a projective plane P, then P−C is called the *complement of two lines in* P.

Denote by $\mathcal{C}$ a class of linear spaces and call its elements $\mathcal{C}$-spaces. Suppose to every $\mathcal{C}$-space L is assigned an order n such that the parameters of L depend only on n. Then every linear space for which there is an integer n such that its parameters can be expressed in the same form in terms of n is called a *pseudo-*$\mathcal{C}$-space (*of order n*).

For example, a pseudo-complement of two lines in a projective plane of order n is a linear space with n^2-n points of degree n+1, n−1 lines of degree n, and n^2 lines of degree n−1. Notice that the definition does not imply that n is the order of a projective plane.

An *inflated affine plane* D consists of an affine plane A together with a linear space L imposed on some of its infinite points. That is, there is a projective plane $P = (p, L)$, a line L of P, a partition $q_1 \cup q_2$ of the points of L, and a linear space $L = (q_1, L')$ such that $D = (p - q_2, L' \cup \{X \cap (p - q_2) \mid X \in L - \{L\}\})$ and $A = P - L$. We also say that D is an affine plane of order n with L *at infinity*. If L is a near-pencil or a projective plane, then D is called a *projectively* inflated affine plane, and if D consists of all the infinite points of A, then it is called a *complete* inflated affine plane. The structure obtained by removing one of the finite points of an inflated affine plane is called an *inflated punctured affine plane*.

There is much more terminology for linear spaces. We do not want to give all definitions, because most explain themselves. For example, a *punctured projective plane* is a linear space which is obtained from a projective plane by removing one of its points, and an *affine plane with an infinite point* is a linear space $A \infty \Pi$ where A is an affine plane and Π is one of its parallel classes.

Now we know most of the linear spaces which occur in this work. There are a few 'exceptional' spaces, which will be denoted by $\mathbf{E}_1, \ldots, \mathbf{E}_9$. Instead of giving the set of points and lines, we define them with the help of a picture. Points are represented by little circles and a line is represented by a (not necessarily straight) line which joins its points. However, we only give a picture of the partial linear space $\mathbf{E}'_j$ obtained by removing the lines of degree 2 from $\mathbf{E}_j$, since $\mathbf{E}_j$ is uniquely determined by $\mathbf{E}'_j$.

$\mathbf{E}_1$: v = 8, b = 11.

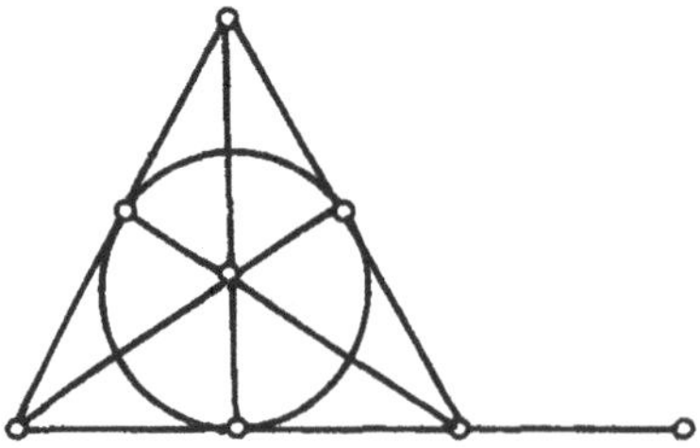

$\mathbf{E}_2$: v = 8, b = 12.

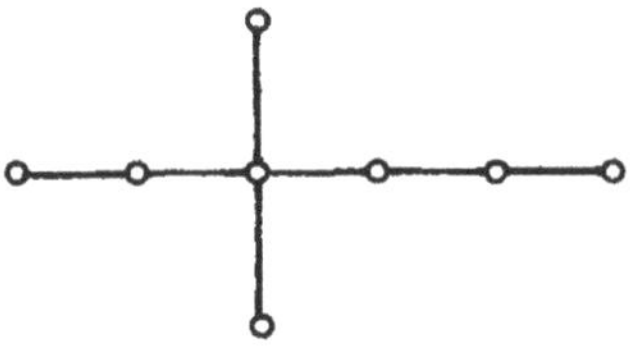

$\mathbf{E_3}$: v = 8, b = 13.

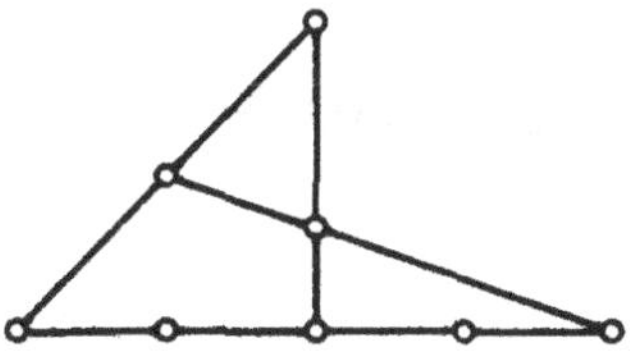

$\mathbf{E_4}$: v = 8, b = 13.

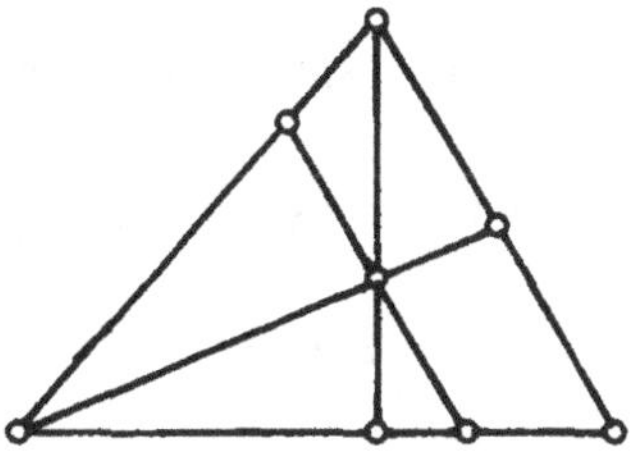

$\mathbf{E_5}$: v = 6, b = 8.

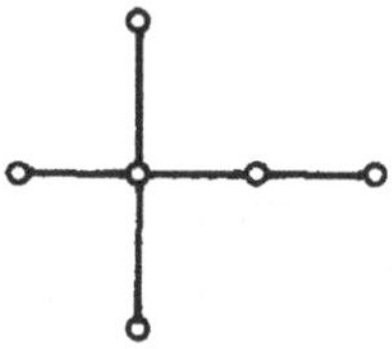

$\mathbf{E_6}$: v = 7, b = 10.

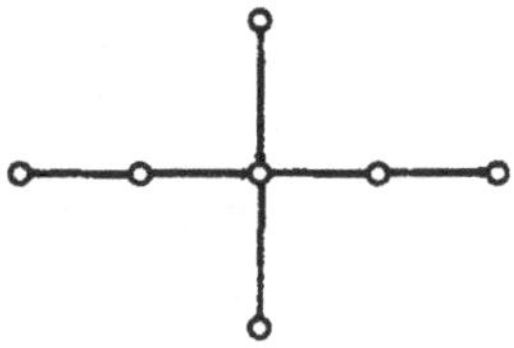

$\mathbf{E_7}$: v = 7, b = 10.

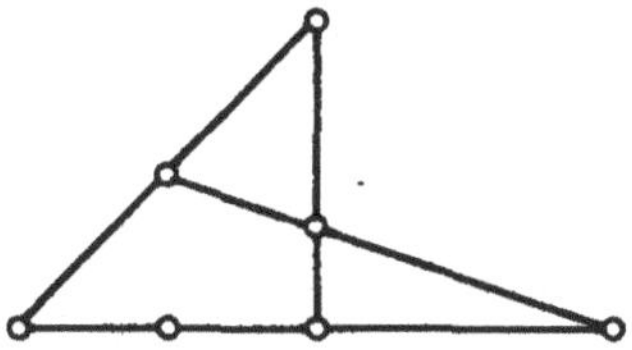

$\mathbf{E_8}$: v = 8, b = 13.

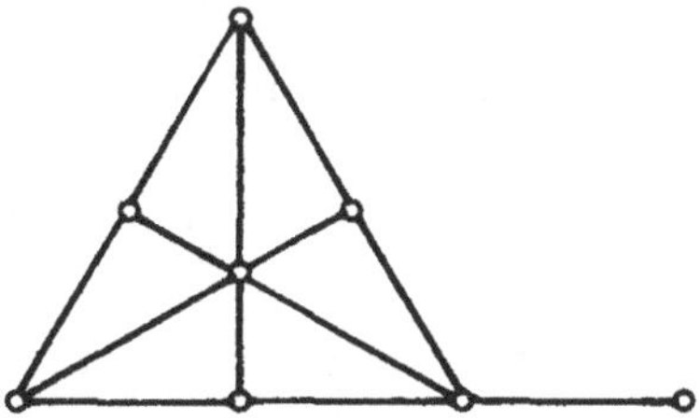

The picture of our last exceptional space $\mathbf{E_9}$ would already be too complicated. It can be defined as follows. Consider the projective plane $P = (p,L)$ of order 4, and let L_1, L_2, L_3 and L_4 be four lines of it which form a quadrilateral. Denote by q the set of the seven points $\neq L_1 \cap L_4, L_2 \cap L_4$ of $L_1 \cup L_2$, and by p_1, p_2, and p_3 the three points $\neq L_1 \cap L_3, L_2 \cap L_3$ of the line L_3. Then we denote the linear space $(p - q, (L - \{L_1, L_2, L_3\}) \cup \{\{p_1,p_2\},\{p_1,p_3\},\{p_2,p_3\}\})$ by $\mathbf{E_9}$. It has 14 points and 21 lines and a unique line of degree 5, which is the line L_4.

Some linear spaces have special names. The near-pencil on three points is called the *triangle*. *Tetrahedron* is another name for the affine plane of order 2, and the *Fano plane* is the projective plane of order 2. The *Fano quasi-plane* is the tetrahedron with the triangle at infinity, which can be obtained from the projective plane of order 2 by *breaking up* one of its lines into three lines of degree 2. The linear space $\mathbf{E}_8$ is called *Lin's Cross.*

Every incidence structure $I = (p, L, I)$ satisfies the following *basic equation*

(B₁) $$\sum_{p\in p} r_p = \sum_{L\in L} k_L.$$

This is true, since both sides of the equation equal to $|I|$. This equation has a lot of important consequences, which will be used throughout this paper.

For every linear space $L = (p, L)$ we have

(B₂) $$v(v-1) = \sum_{L\in L} k_L(k_L-1).$$

This follows from the basic equation for the incidence structure whose set of points is $P = \{(p,q) \mid p$ and q are distinct points of $L\}$, whose lines are the sets $G_L = \{(p,q) \in P \mid p,q \in L\}$, $L \in L$, and in which a point (p,q) lies on a line G_L if $p,q \in L$. Equation (B₂) reflects the fact that any two points of a linear space lie on a unique line.

If p is a point of a linear space (p, L), then the equation (B₁) used for the partial linear space $(p-\{p\}, \{L \in L \mid p \in L\})$ shows

(B₃) $$v-1 = \sum_{p\in L} (k_L-1).$$

This equation reflects the fact that the point p is joined to every other point by a unique line.

Let L be a line of a linear space $L = (p, L)$ and denote by M the set of lines which are parallel to L. The basic equation (B₁) used for $(p-L, M)$ gives

(B₄) $$\sum_{p\notin L} (r_p-k_L) = \sum_{L\in M} k_L,$$

since every point p outside of L lies in r_p-k_L lines parallel to L.

Obviously, the equations (B₁), (B₂), (B₃), and (B₄) also hold for $(n+1,1)$-designs.

Now we prove two easy lemmas, which will be used frequently throughout this paper. We call them *Transfer-Lemma* and *Parallel-Lemma.*

TRANSFER-LEMMA. Let $\mathbf{L} = (p, L, I)$ be an incidence structure with the property that any two points lie on a unique line.

a) Suppose that L and L' are lines and that p is a point outside of L and L'. If t is the number of lines through p which meet L and miss L', then $t+k_{L'}-k_L$ is the number of lines through p which meet L' and miss L.

b) Suppose that L, L', and H are lines and that H meets L and misses L'. Then there are at least $(k_H-1)(1+k_{L'}-k_L)$ lines which meet H and L' and which miss L.

Proof. a) The point p lies on k_L-t lines which meet L and L', and it lies on $k_{L'}$ lines which meet L'. Therefore p lies on $k_{L'}-(k_L-t)$ lines which meet L' and miss L.

b) By the first part of the lemma each point $\neq H\cap L$ of H lies on at least $1+k_{L'}-k_L$ lines which meet L' and miss L. The assertion follows.■

PARALLEL-LEMMA. Let $\mathbf{L} = (p, L, I)$ be an incidence structure with the property that any two points lie on a unique line. Suppose that N is a line of degree n, and that L and L' are two intersecting lines which are parallel to N. Denote the degree of L_j by $n+1-d_j$ and set $t = \sum_{p\in N} (r_p-n-1)$. Then $d_1 \cdot d_2 \geq n-t$.

Proof. If b is the number of lines, then there are $m := b-1-n^2-t$ lines which miss N. Let q be the point of intersection of L_1 and L_2, denote the degree of q by $n+1+t_q$, and set $t_j = \Sigma(r_p-n-1)$ where the sum runs over all points $p \neq q$ of L_j. Then q lies on $1+t_q$ lines which miss N, and L_j meets t_j lines in a point $p \neq q$ which miss N. Hence, there are at least $m-1-t_q-t_1-t_2$ lines which miss N, L_1, and L_2, which implies that there are at least $m-t_q-t_1-t_2$ lines which miss L_1 and L_2.

Now we determine the exact number of lines which miss L_1 and L_2. Using the definition of t_1 and t_q, we see that the line L_1 meets $(n+1-d_1)n+t_1+t_q$ other lines. Furthermore, the line L_2 meets $(n-d_2)d_1+t_2$ lines which miss L_1. Hence, the number of lines which miss L_1 and L_2 is $b-1-[(n+1-d_1)n+t_1+t_q]-[(n-d_2)d_1+t_2] = b-n^2-n-1+d_1d_2-t_1-t_2-t_q$. It follows that $b-n^2-n-1+d_1d_2-t_1-t_2-t_q \geq m-t_q-t_1-t_2$. In view of $m = b-1-n^2-t$, the assertion follows.■

Especially the Parallel-Lemma will be extremly helpful in many proofs. It will be used to construct sets of mutually disjoint lines containing an n-line.

We define a function $B(v)$ for integers $v \geq 4$ in the following way. Let n be the unique positive integer with $(n-1)^2+(n-1)+1 < v \leq n^2+n+1$. Then

$$B(v) = \begin{cases} n^2+n-1, & \text{if } v = n^2-n+2 \text{ and } v \neq 4 \\ n^2+n, & \text{if } n^2-n+3 \leq v \leq n^2+1 \text{ or } v = 4 \\ n^2+n+1, & \text{if } n^2+2 \leq v \end{cases}$$

We shall see in section 2 that every non-degenerate linear space with v points and b lines satisfies not only $b \geq v$ but also $b \geq B(v)$.

2. Lower bounds for the number of lines

We begin this section with a proof of the famous theorem of de Bruijn and Erdős (1948). It says that every linear space has at least as many lines as points and that the generalized projective planes are characterized by $v = b$. There are a lot of different proofs for this result, see e.g. Hanani (1951, 1954–1955), Bouten and de Witte (1965), and de Witte (1975b). We shall also prove a little stronger property, which will be very helpful in the proof of Totten's theorem.

One of the most fascinating proofs for the Theorem of de Bruijn and Erdős is due to Conway (see Basterfield and Kelly, 1968). Suppose that $L = (p, L)$ is a linear space with v points and b lines which satisfies $b \le v$. Consider any non-incident point line pair (p,L). Then $r_p \ge k_L$ with equality if and only every line through p meets L. Hence

$$\frac{r_p}{b-r_p} \ge \frac{k_L}{v-k_L}$$

Now we take the sum over all non-incident point line pairs. Because each point p occurs in $b-r_p$ such pairs and because each line occurs in $v-k_L$ such pairs, we conclude

$$\sum_{p\in p} r_p \ge \sum_{\substack{p,L\\ p\notin L}} \frac{r_p}{b-r_p} \ge \sum_{\substack{p,L\\ p\notin L}} \frac{k_L}{v-k_L} = \sum_{L\in L} k_L$$

In view of the basic equation (B_1) in section 1, we obtain equality. Hence $v = b$ and $r_p = k_L$ holds for every non-incident point line pair. This implies that any two lines intersect so that L is a generalized projective plane.

We shall present a more complicated proof, which has the advantage that it can be used to prove much stronger results. The idea to use incidence matrices, goes probably back to Majumdar (1953).

2.1 Lemma. Suppose that $A = (a_{jk})$ is an $n \times n$ matrix and that there exists a partition $I_1 \cup \ldots \cup I_t$ of $\{1,\ldots,n\}$ such that $a_{jj} \ne 0$ for all j, $a_{jk} = 0$ for distinct j,k of the same set I_s, and else $a_{jk} = 1$. For $k \in \{1,\ldots,t\}$ set $s_k = \sum_{j\in I_k} \frac{1}{a_{jj}}$. If $0 \le s_k \le 1$ for all k, and $s_k = 1$ for at most one value of k, then A is regular.

Proof. Suppose that $A(x_j) = 0$ with indeterminates x_j. Define

$$y_k = \sum_{j=1}^{n} x_j - \sum_{j \in I_k} x_j, \qquad k = 1,\dots,t,$$

and the matrix

$$B = \begin{pmatrix} 1 & s_2 & s_3 & \dots & & s_t \\ s_1 & 1 & s_3 & \dots & & s_t \\ s_1 & s_2 & 1 & & & . \\ . & & & . & & . \\ . & & & & . & . \\ . & & & & 1 & s_t \\ s_1 & & \dots & & s_{t-1} & 1 \end{pmatrix}.$$

In view of $A(x_j) = 0$, we have

(+) $\quad a_{jj}x_j + y_k = 0,$ for $k = 1,\dots,t$ and all $j \in I_k$,

and thus

$$y_k = \sum_{h \neq k} \sum_{j \in I_h} x_j = \sum_{h \neq k} \sum_{j \in I_h} \frac{-y_h}{a_{jj}} = -\sum_{h \neq k} y_h s_h,$$

i.e. $B(y_j) = 0$.

Since by hypothesis, $0 \leq s_k \leq 1$ and $s_k = 1$ for at most one k, the matrix B is regular. Thus $y_j = 0$ for all j. In view of (+) this implies that $(x_j) = 0$, so A is regular.■

2.2 Lemma. Let q be a point of a linear space L, and denote by $L_1,\dots,L_r$ the lines through q such that $|L_j| \leq |L_r|$ for all j. We set

$$S_k = \sum_{q \neq p \in L_k} \frac{1}{r_p - 1}$$

and denote by w the number of indices $j < r$ with $S_j \neq 1$. Then we have:

a) If $j < r$, then $0 < S_j \leq 1$. We have $S_j = 1$ if and only if $|L_j| = |L_r|$ and $r_q = |L_r|$ for all points $p \in L_j - \{q\}$.

b) If $w \leq r-2$, then $b \geq v+w$.

c) If $w = r-1$, then $b \geq v+r-2$. If $w = r-1$ and $S_r \leq 1$, then $b \geq v+r-1$.

Proof. a) Trivially $S_j > 0$ for all $j = 1,\dots,r$. If $j < r$, then every point $\neq q$ of L_j has degree at least $|L_r| \geq |L_j|$, so $S_j \leq 1$ with equality if and only if $|L_j| = |L_r|$ and $r_p = |L_j|$ for all points p of $L_j - \{q\}$.

b) We may assume that $S_k < 1$ for $k \leq w$ and $S_k = 1$ for $w < k < r$.

Denote the points $\neq q$ by $p_1,\ldots,p_{v-1}$. Let L' be the incidence structure which is obtained from L by removing the point q and the lines $L_1,\ldots.,L_{w+1}$, and let C be an incidence matrix of L' in which the j-th row corresponds to the point p_j, $j = 1,\ldots,v-1$. Furthermore, let $A = (a_{jk})$ be the matrix CC^t and let $\cup I_k$ be the partition of $\{1,\ldots,v-1\}$ with $I_k = \{j \mid p_j \in L_k\}$, $k = 1,\ldots,w+1$, and $|I_k| = 1$ for $k > w+1$. Then a_{jj} is the degree of p_j in L'. Also, a_{jk}, $j \neq k$, is the number of lines of L' which pass through p_j and p_k. Hence $a_{jk} = 0$ if the line p_jp_k of L is one of the lines $L_1,\ldots,L_{w+1}$ and $a_{jk} = 1$ if p_jp_k is a line of L'. Define s_k as in Lemma 2.1. Then $0 \le s_k = S_k < 1$ for $k \in \{1,\ldots,w\}$ and $s_{w+1} = S_{w+1} = 1$. If $k \ge w+2$, then $I_k = \{j\}$ for some point $p = p_j \notin L_1\cup\ldots\cup L_{w+1}$; in this case $s_k r_p = 1$ so that $0 < s_k < 1$. It follows therefore from Lemma 2.1 that A is regular. Since $A = CC^t$, this implies that the rank of the $(v-1)x(b-w-1)$ matrix C is $v-1$. Consequently, $v-1 \le b-1-w$.

c) If $w = r-1$ and $S_r \le 1$, then $b \ge v+w$ follows similarly (this time, let L' be the incidence structure obtained from L by removing p and all the lines L_j). By Lemma 2.1, the obtained matrix A is still regular. However, this argument fails, if $S_r > 1$. In this case, we can only remove the lines $L_1,\ldots,L_{r-1}$. C is then a $(v-1)x(b+1-r)$ matrix so that we obtain only $b \ge v+r-2$.■

<u>2.3 Theorem</u> (De Bruijn and Erdös, 1948). Every linear space L satisfies $b \ge v$ with equality if and only if L is a generalized projective plane.

<u>Proof</u>. By Lemma 2.2, $b \ge v$. Let us suppose $b = v$. We have to show that L is a near-pencil or a projective plane. Obviously, L is a near-pencil if L has a point of degree two so that we may assume that every point has degree at least three.

Let q be any point of L and use the notation of Lemma 2.2. Since $r = r_q \ge 3$, part b) and c) of Lemma 2.2 show that $w = 0$, i.e. $s_j = 1$ for $j < r$. Part a) shows therefore that every line through q and every point outside L_r has degree $x := |L_r|$. Now every line through q has the same degree r, and we may consider any order of the lines through q. This implies that every point other than q has degree x. Since q was an arbitrary point, we conclude that every point or line of L has degree x so that L is a projective plane of order $x-1$.■

We now want to prove a much better bound for b. A very important aid will be the following result. To state it, we define the function f which will be used throughout:

$$f(k,v) := 1 + \frac{k^2(v-k)}{v-1} \quad \text{for all real numbers } k \text{ and } v \text{ with } 2 \le k \le v.$$

2.4 Lemma (Stanton and Kalbfleisch, 1972). Let $\mathbf{L}$ be a linear space and denote its number of points and lines by v and b.

a) If k is the degree of some line L, then $b \ge f(k,v)$ with equality if and only if L meets every other line and if every line other than L has the same degree k', which is given by $k(k'-1) = v-1$ in this case.

b) Suppose that k_1 and k_2 are real numbers satisfying $k_1 \le k_2$, $b < f(k_1,v)$ and $b < f(k_2,v)$. Then $k_L < k_1$ or $k_L > k_2$ for every line L of $\mathbf{L}$.

Proof. a) Let $\mathcal{M}$ be the set of lines which intersect L, $m := |\mathcal{M}|$. Then we have

$$\sum_{L\in\mathcal{M}} 1 = m, \qquad \sum_{L\in\mathcal{M}} (k_L-1) = k(v-k), \qquad \text{and}$$

(1) $$\sum_{L\in\mathcal{M}} (k_L-1)(k_L-2) \le (v-k)(v-k-1).$$

The third equation follows from the fact that any two points outside of L are connected by at most one line of $\mathcal{M}$. We obtain

$$\sum_{L\in\mathcal{M}} k_L = k(v-k)+m$$

and

$$\sum_{L\in\mathcal{M}} k_L^2 \le (v-k)(v+2k)+m.$$

Substituting these relations in the Cauchy-Schwarz inequality

(2) $$\left(\sum_{L\in\mathcal{M}} k_L\right)^2 \le m\left(\sum_{L\in\mathcal{M}} k_L^2\right)$$

shows that $m \ge f(k,v)-1$ with equality if and only if equality holds in (2), which means that every line of $\mathcal{M}$ has the same degree, and in (1), which means that any two points outside of L are connected by a line of $\mathcal{M}$ or equivalently that L meets every other line. Hence, $b \ge m+1 \ge f(k,v)$ with equality if and only if every line other than L meets L and has the same degree k'.

Now suppose $b = f(k,v)$ and let p be a point outside of L. Since every line through p meets L and has degree k', we have $r_p = k$ and $v-1 = r_p(k'-1) = k(k'-1)$.

b) The definition of f implies that $f(k,v) \ge \max\{f(k_1,v), f(k_2,v)\}$ for all numbers k, k_1, k_2 with $k_1 \le k \le k_2$. Part b) follows therefore immediately from part a).■

2.5 Lemma. Let L be a non-degenerate linear space satisfying $n^2-n+2 \le v$ and $b \le n^2+n+1$ for some integer $n \ge 2$, and denote by k the maximal line degree. If $k > n+1$, then L is one of the linear spaces $\mathbf{E}_2$ and $\mathbf{E}_3$.

Proof. Since L is non-degenerate, we have $n+2 \le k \le v-2$. From 2.4, we obtain

$$n^2+n+1 \ \ge\ b \ \ge\ \min\{f(v-2,v),f(n+2,v)\} \ =\ f(n+2,v) \ \ge\ f(n+2,n^2-n+2)$$
$$= n^2+n+1 \ +\ \frac{2n^3-4n^2-9n}{n^2-n+1}\ .$$

This shows that $n \le 3$. In view of $b \ge f(k,v) \ge f(k,k+2) > 2k-1 \ge 2n+3$, we conclude that $n = 3$ and $k = 5$ or $k = 6$. Now $13 \ge b \ge v \ge 8$ so that $b \ge f(k,v)$ implies $v = 8$. It is now easy to check that $L = \mathbf{E}_2$, if $k = 6$, and $L = \mathbf{E}_3$, if $k = 5$.■

2.6 Theorem (Erdős, Mullin, Sós, and Stinson, 1983). Every non-degenerate linear space satisfies $b \ge B(v)$.

Proof. Let n be the unique integer with $n^2-n+2 \le v \le n^2+n+1$, and let k denote the maximal line degree. If $k \ge n+2$, then Theorem 2.6 follows from Lemma 2.5. We may therefore assume that $k \le n+1$. In view of $v \ge n^2-n+2$, this implies that every point has degree at least n. We consider three cases.

Case 1. There is a line L of degree n+1.
If every point of L has degree at least n+1, then $b \ge k_L \cdot n+1 \ge B(v)$. Assume therefore that L has a point p of degree n.

If p is contained in a second line H of degree n+1, then every point $\ne p$ of L has degree at least k_H so that L meets at least n^2+n-1 other lines. Thus $b \ge n^2+n$ and $v \le 1+r_p \cdot n = n^2+1$, so $b \ge B(v)$.

If L is the only line of degree n+1 through p, then $v = n^2-n+2$, and p is contained in L and n−1 lines $N_2,\dots,N_n$ of degree n. Since L is not a near-pencil, $n \ge 2$. Since every point $\ne p$ of $N_2 \cup N_3$ has degree at least $k_L = n+1$, there are at least n^2+n-1 lines which meet N_2 or N_3. Hence $b \ge n^2+n-1 = B(v)$.

Case 2. The maximal line degree is n.
We have $v-1 \le r_p(n-1)$ and thus $r_p \ge n+1$ for every point p. A line N of degree n meets therefore at least n^2 other lines. Hence, $b \ge n^2+1+c$, where c is the number of lines parallel to N. Since each point outside of N lies on at least one line parallel to N, $v-k_N \le cn$ (see equation (B_4) of section 1), and thus $b \ge B(v)$.

Case 3. Every line has degree at most $n-1$.

As in the last case, every point has degree at least $n+1$ so that $v(n+1) \leq b(n-1)$ (this follows from the basic equation (B_1) for incidence structures) and therefore $b \geq n^2+n+1 \geq B(v)$.■

Theorem 2.6 considerably improves the bound $b \geq v$ from Theorem 2.3. It does not, however, generalize the equality part in the De Bruijn-Erdös theorem. Such a 'full' generalization could read as follows: Every linear space L with $n^2-n+2 \leq v \leq n^2+n+1$ satisfies $b \geq B(v)$ and equality implies that L can be embedded into a projective plane of order n. This generalization is almost true. We shall show later that the linear space E_1 is the only exception.

Special cases of Theorem 2.6 are already known for a longer time. For example, Bruen (1973) proved in 1973 that every non-degenerate linear space with n^2 points has at least n^2+n lines with equality if and only if it is an affine plane of order n, or a punctured affine plane of order n with one point at infinity.

Theorem 2.3 says that every linear space has as many lines as points. This result gave the idea to study linear spaces which have only a few more lines than points. This idea was first followed by Bridges (1972) who classified the linear spaces with $b = v+1$ in 1972. Almost throughout in this work, we shall study linear spaces with "few lines" in this sense. However, there are other points of view to see "linear spaces with few lines".

Hanani (1954-55) and Varga (1985) proved that a line of maximal degree of a linear space meets at least $v-1$ other lines with equality if and only if the space is a generalized projective plane (this improves the result of de Bruijn and Erdös!). A linear space with few lines may therefore be seen as a linear space in which a line of maximal degree meets at most $v-1+c$ other lines for small values of c. We shall consider this problem in section 15.

Another result in this area is the proof of the Dowling-Wilson Conjecture, which we present in section 17: If some point of a linear space lies on t lines which miss a given line, then $b \geq v+t$.

3. Basic properties and results of (n+1,1)-designs

The purpose of this work is to study linear spaces with almost "the same" number of points and lines. If n is the unique positive integer satisfying $(n-1)^2+(n-1)+1 < v \leq n^2+n+1$, we shall see in the next sections that most of the v points of a linear space L have degree n+1. We therefore obtain an (n+1,1)-design D if we remove the points of degree $\neq$ n+1. Information about (n+1,1)-designs are then very helpful for the characterization of L. For example if we know that D can be embedded into a projective plane of order n, it is sometimes very easy to see that also L can be embedded. In this section, some basic properties and results of (n+1,1)-designs are developed.

For every (n+1,1)-design D we set $k_L = n+1-d_L$ for every line L of D. We recall that two lines are called parallel if they are distinct and disjoint.

<u>3.1 Lemma</u>. Let D be an (n+1,1)-design, denote its number of points by $v = n^2+n+1-s$ and its number of lines by $b = n^2+n+1+z$. Then

a) The number of lines parallel to a line L is $d_L \cdot n+z$.

b) The number of lines parallel to two lines L and H is $d_H \cdot d_L+z$, if H and L intersect, and it is $n-1+(d_H-1)(d_L-1)+z$, if H and L are parallel.

c) If $\mathcal{M}$ is the set of lines parallel to a given line L, then $(v-k_L)d_L = \sum_{X\in\mathcal{M}} k_X$.

<u>Proof</u>. a) follows from $b = n^2+n+1+z$ and the fact that L meets $k_L \cdot n$ other lines.

b) In both cases it is easy to determine the number of lines which miss L and meet H; b) then follows from a).

c) is true, since each of the n^2-s+d_L points outside of L lies on exactly d_L of the lines which are parallel to L (this follows also from equation (B_4) defined in chapter 1).■

<u>3.2 Lemma</u>. Every (n+1,1)-design $D = (\mathcal{p},\mathcal{L})$ with $n^2+n+1-s$ points and $n^2+n+1+z$ lines satisfies

$$\sum_{L\in\mathcal{L}} d_L = (s+z)(n+1) \quad \text{and} \quad \sum_{L\in\mathcal{L}} d_L^2 = s(n+s)+(n+1)^2z.$$

Proof. Consider the equations (B_2) and (B_1) in chapter 1. (B_2) says that $v(v-1) = \sum k_L(k_L-1)$, and, since every point has degree n+1, (B_1) says that $\sum k_L = v(n+1)$. Substituting $k_L = n+1-d_L$, the two desired equations can be obtained from these two equations.■

3.3 Lemma. Suppose that D is an (n+1,1)-design with $n^2+n+1-s$ points and n^2+n+1 lines.

a) If b_n is the number of lines of degree n, then $b_n \geq s(n+2-s)$. In particular, if $n^2+1 \leq v \leq n^2+n-1$, then $b_n \geq 2n$.

b) If there is no line of degree n, then $\sum_{L \in \mathcal{L}} d_L(d_L-2) = s(s-2-n)$.

Proof. Lemma 3.2 shows that

$$\sum_{L \in \mathcal{L}, k_L \neq n} d_L(d_L-2) = s(s-2-n) + b_n,$$

and this proves a) and b).■

Now we prove our first embedding result. The technique which is used is typical. Hence we describe this process in some detail.

3.4 Lemma. Suppose that an (n+1,1)-design $D = (\mathcal{p},\mathcal{L})$ with n^2+n+1 lines and v points has t mutually intersecting lines of degree n. Then D can be embedded into an (n+1,1)-design with n^2+n+1 lines and v+t points.

Proof. Let $N_1,\dots,N_t$ be t mutually intersecting lines of degree n. Since each point outside a line N_j has degree n+1, each point outside N_j is contained in a unique line missing N_j. Hence, the set π_j consisting of N_j and the lines parallel to N_j is a parallel class of D, $j = 1,\dots,t$. By Lemma 3.1a), $|\pi_j| = n+1$. For $j \neq k$, the lines N_j and N_k intersect so that N_j intersects n of the lines of π_k, i.e. π_j and π_k have a unique line in common. Now we extend D in such a way that the lines of π_j meet in an infinite point. In order to do so, let ∞_j, $j = 1,\dots,t$, be new symbols and define the incidence structure $D' = (\mathcal{p}\cup\{\infty_1,\dots,\infty_t\},\mathcal{L},I)$ in the following way:

For all $p \in \mathcal{p}$ and $L \in \mathcal{L}$,	$p\ I\ L$	if and only of if $p \in L$.
For all $L \in \mathcal{L}$,	$\infty_j\ I\ L$	if and only if $L \in \pi_j$.

We show that any two points p and q of D' are contained in a unique line of D'. Indeed, if $p = \infty_j$ and $q = \infty_k$ are infinite points, then the line of $\Pi_j \cap \Pi_k$ is the unique line which contains p and q. If p is a finite point, i.e. a point of D, and $q = \infty_j$ an infinite point, then the unique line of the parallel class Π_j of D which contains p is the only line containing p and ∞_j in D'. If p and q are points of D, then the line pq exists and is unique, because D is an (n+1,1)-design and because we have not adjoined lines to D or deleted lines from D. Hence, any two points of D' are contained in a unique line. In view of $|\Pi_j| = n+1$, it follows that D' is an (n+1,1)-design with v+t points, which is an extension of D.■

3.5 Corollary. a) (Vanstone, 1973) Every (n+1,1)-design D with n^2+n+1 lines and $v \geq n^2$ points can be embedded into a projective plane of order n.

b) (De Witte, 1977b) Suppose that an (n+1,1)-design L has n^2+n+1 lines and that some point p is contained in t lines of degree n. If $v+t \geq n^2$, then D can be embedded into a projective plane of order n.

Proof. a) We proceed by induction on $s = n^2+n+1-v$. If $s = 0$, then Lemma 3.2 shows that every line of D has degree n+1 so that D itself is a projective plane.

Let us now suppose that $s > 0$. Lemma 3.1 a) shows that D has a line N of degree n. By Lemma 3.4, D can be extended so that the induction hypothesis shows that D can be embedded into a projective plane of order n.

b) follows from Lemma 3.4 and part a).■

We shall show in chapter 7 that $v > n^2-\frac{1}{2}n+6$ and $b = n^2+n+1$ imply embeddability. This is a much more difficult problem. However, if there are at most n^2+n lines, then the bound for v can be improved very easily.

3.6 Lemma. Every (n+1,1)-design D with at least n^2-n+1 points and at most n^2+n lines can be embedded into a projective plane of order n.

Proof. Lemma 3.1 a) shows that every line has degree at most n. Suppose that N is an n-line. Since every point outside of N has degree n+1, N together with the lines parallel to N forms a parallel class Π. Since every point of N has degree n+1, we have $b = n^2+|\Pi|$ so that Π has at most n elements. On the other hand, since every line of Π has degree at most n, we have $|\Pi|n \geq v$ so that $|\Pi| = n$

and $b = n^2+n$. We want to show that there are n+1 mutually disjoint parallel classes Π with $|\Pi| = n$.

Suppose we have already constructed w mutually disjoint parallel classes $\Pi_1,\ldots,\Pi_w$ with n elements, $0 \le w \le n$. If $\mathcal{M}$ is the set of lines not contained in any of the parallel classes Π_j, then $|\mathcal{M}| = b-wn = (n+1-w)n$. Counting the set

$$\{(p,L) \mid L \text{ is a line of } \mathcal{M} \text{ and } p \text{ is a point of } L\}$$

in two different ways, we obtain

$$v(n+1-w) = \sum_{L\in \mathcal{M}} k_L.$$

Since $|\mathcal{M}|(n-1) = (n+1-w)n(n-1) < v(n+1-w)$, this shows that some line N of $\mathcal{M}$ has degree n. Let Π be the parallel class consisting of N and the lines parallel to N. Then $|\Pi| = n$. Furthermore Π is disjoint to Π_j, $j \le w$, since the n-line N intersects each line of the parallel-class Π_j.

Consequently, there are n+1 mutually disjoint parallel classes $\Pi_1,\ldots,\Pi_{n+1}$. As in Lemma 3.4, we add an infinite point ∞_j to each line of Π_j. But this time, the resulting structure is not an (n+1,1)-design, since the infinite points have degree n and since there is no line passing through any two of the infinite points. However, if we adjoin also the new line $\{\infty_1,\ldots,\infty_{n+1}\}$, then we obtain an (n+1,1)-design. This design has v+n+1 points and n^2+n+1 lines. By our last lemma, it can be embedded into a projective plane of order n, and therefore also D can then be embedded into this projective plane.■

We have seen that parallel classes play a crucial role for the embedding problem. They have been constructed very easily with the help of lines of degree n. But what can be done when there are no lines of degree n? Consider for example an (n+1,1)-design D with n^2+n+1 lines and n^2-1 points. Then Lemma 3.3b) does not exclude the possibility that every line has degree n−1 or n+1, and in fact such designs exist. A line L of degree n−1 is then parallel to 2n lines and every point outside L is contained in two lines parallel to L. If D can be embedded into a projective plane of order n, then the set $\mathcal{M}$ of the 2n lines parallel to L can be partitioned into sets $\mathcal{M}_1$ and $\mathcal{M}_2$ such that $\{L\} \cup \mathcal{M}_1$ and $\{L\} \cup \mathcal{M}_2$ are parallel classes of D. In this case, there do not exist three mutually intersecting lines in M. On the other hand, in order to show that a partition $M = M_1 \cup M_2$ exists, it would be extremely helpful to know that M does not contain three mutually intersecting lines. More generally, a line of degree n+1−d should not miss d+1 lines each of which intersects the d other ones. The following lemma gives a useful criterion for this. The method used to prove this lemma is due to Bruck (1963) and Bose (1963).

<u>3.7 Lemma</u>. Let D be an (n+1,1)-design and denote its number of lines by $n^2+n+1+z$. Suppose that a line L is parallel to r mutually intersecting lines $L_1,\dots,L_r$. Denote the degree of L by $n+1-d$ and the degree of L_j by $n+1-d_j$.

a) If $r = d+1$, then

$$n \le \sum_{j<k} d_j d_k - (d-1)\sum_j d_j + \tfrac{1}{2}(d-2)(d+1) + \tfrac{1}{2}d(d-1)z.$$

b) If $r < d$, and if t is the number of lines which are parallel to L and intersect $L_1,\dots,L_r$, then

$$t \ge (d-r)n - (d-1)\sum_{j\le r}(d_j-1) - z(r-1).$$

<u>Proof</u>. Let T denote the set of the $dn-r+z$ lines $\ne L_j$ which are parallel to L, and for $0 \le y \le r$ denote by $f(y)$ the number of lines in T which are parallel to exactly y of the lines L_j. If we denote for any lines $H_1,H_2,\dots,H_c$ the number of lines which are parallel to all H_j by $m(H_1,\dots,H_c)$, then we have

$$\sum_{y=0}^{r} f(y) = |T| = dn-r+z$$

$$\sum_{y=1}^{r} f(y)y = \sum_{j=1}^{r} m(L,L_j) = \sum_{j=1}^{r} [n-1+(d-1)(d_j-1)+z], \quad \text{and}$$

$$\sum_{y=2}^{r} f(y)y(y-1) = \sum_{j\ne k} m(L,L_j,L_k) \le \sum_{j\ne k} (m(L_j,L_k)-1) \le \sum_{j\ne k} (d_jd_k-1+z).$$

If $r = d+1$, then it follows

$$0 \le \frac{1}{2}\sum_{y=0}^{r} f(y)(y-1)(y-2) = \sum_{y=0}^{r} f(y) - \sum_{y=0}^{r} f(y)y + \frac{1}{2}\sum_{y=0}^{r} f(y)y(y-1)$$

$$\le dn-d-1+z - \sum_{y=0}^{d+1} [n-1+(d-1)(d_j-1)+z] + \sum_{j<k} (d_jd_k-1+z),$$

which implies a). b) follows from $t = f(0) \ge \sum_{y=0}^{r} f(y) - \sum_{y=0}^{r} f(y)y$.■

We conclude this chapter with a lemma which shows that embeddings can sometimes be extended.

<u>3.8 Lemma</u>. Let $\mathbf{L} = (p,\mathcal{L})$ be an $(n+1,1)$-design with $b = n^2+n+1$ lines. Suppose that there is a set $\mathcal{L}' \subseteq \mathcal{L}$ of at least n^2+1 lines such that the structure $\mathbf{L}' = (p,\mathcal{L}')$, which is induced by the lines of $\mathcal{L}'$ on p, can be embedded into a projective plane $\mathbf{P} = (p_0,\mathcal{L}_0)$ of order n. If $\mathbf{L}$ has at least n^2-n+2 points, then also $\mathbf{L}$ is embeddable into $\mathbf{P}$.

<u>Proof</u>. We proceed by induction on $t = |\mathcal{L}| - |\mathcal{L}'| = n^2+n+1-|\mathcal{L}'|$. If $t = 0$, then $\mathbf{L} = \mathbf{L}'$ so nothing is to show. Consider the case $t \geq 1$. We have to show that there is a line L in $\mathcal{L}-\mathcal{L}'$ and a line L_0 in $\mathcal{L}_0-\mathcal{L}'$ such that every point of $L \subseteq p$ is also a point of L_0; then $(p,\mathcal{L}' \cup \{L\})$ is embedded in $\mathbf{P}$ and by the induction hypothesis also $\mathbf{L}$ is embedded in $\mathbf{P}$.

Consider first the case that $\mathbf{L}'$ has a point p of degree n and denote by L the unique line of $\mathcal{L}-\mathcal{L}'$ through p. Then there is also a unique line L_0 of $\mathbf{P}$ which passes through p and which is not a line of $\mathbf{L}'$. Since p is connected to every point outside of L by a line of $\mathcal{L}'$, every point of L is contained in L_0.

Suppose now that no point of $\mathbf{L}'$ has degree n and consider the partial linear space $\mathbf{D} = (p_0,\mathcal{L}')$ which is obtained from $\mathbf{P}$ by removing the lines which are not lines of $\mathbf{L}'$ (here we consider each line of $\mathcal{L}'$ as a line of $\mathbf{P}$). Then $\mathbf{L}'$ lies in $\mathbf{D}$. $\mathbf{D}$ has n^2+n+1 points and constant line degree $n+1$. Furthermore, any two distinct lines of $\mathbf{D}$ intersect in $\mathbf{D}$. The structure which is dual to $\mathbf{D}$ is therefore a linear space with constant point degree $n+1$, n^2+n+1 lines and $|\mathcal{L}'|$ points. In view of $|\mathcal{L}'| = n^2+n+1-t$, Lemma 3.3 a) shows that $\mathbf{D}$ has at least $t(n+1-t)$ points of degree n. Since $\mathbf{L}'$ has at least $n^2+n+1-(2n-1)$ points, each of which is also a point of $\mathbf{D}$ and has in $\mathbf{D}$ the same degree as in $\mathbf{L}'$, and because $\mathbf{L}'$ has no point of degree n, we conclude $t(n+1-t) \leq 2n-1$, so $t = 1$ in view of $1 \leq t \leq n$. Let L be the line of $\mathbf{L}$ with $\mathcal{L} = \mathcal{L}' \cup \{L\}$ and let L_0 be the line of $\mathbf{P}$ with $\mathcal{L}_0 = \mathcal{L}' \cup \{L_0\}$. L has degree 0 in $\mathbf{L}$ (in view of $\mathcal{L} = \mathcal{L}' \cup \{L\}$ every point of L would have degree n in $\mathbf{L}'$), so $L \subseteq L_0$.■

4. Points of degree n.

The study of a linear space **L** with $n^2-n+2 \le v \le b \le n^2+n+1$ begins with the determination of the degrees of the points and lines. We have already shown that **L** is a near-pencil, $\mathbf{E}_2$, or $\mathbf{E}_3$, if some line has degree at least n+2. Now we examine **L**, if some point has degree at most n. We shall succeed to determine **L** unless $v = n^2-n+2$ and $b = n^2+n+1$.

<u>4.1 Lemma</u>. Suppose the linear space **L** has a point q of degree at most n and satisfies $n^2-n+2 \le v \le b \le n^2+n$. If **L** has a line L of degree n+1 which contains a point q of degree n and n points of degree n+1, then **L** can be embedded into a projective plane of order n.

<u>Proof</u>. The line L meets n^2+n-1 other lines. Because of $b \le n^2+n$, this shows that $b = n^2+n$ and that L meets every other line. In particular, every point outside of L has degree $k_L = n+1$. Consequently, if we remove the point q from **L**, then we obtain an (n+1,1)-design **D**, which can be embedded into a projective plane **P** of order n (Lemma 3.6). Because L has degree n in **D**, there is a unique point q' in **P**−**L** which lies on L. Since L meets every line in **P**, each line which is parallel to L in **D** passes in **P** through q'. This shows that the line of **D** which pass in **P** through q' are the lines which pass in **L** through q. It follows that **L** can be embedded into a projective plane which is isomorphic to **P**.■

<u>4.2 Lemma</u>. Suppose that **L** is a non-degenerate linear space satisfying $n^2-n+2 \le v \le b \le n^2+n$. If **L** has a point q of degree at most n, then one of the following cases occurs.

(1) **L** can be embedded into a projective plane of order n.

(2) **L** is one of the linear spaces $\mathbf{E}_1$ or $\mathbf{E}_2$.

(3) $v = n^2-n+2$ and $b = n^2+n$. The point q lies on one line L of degree n+1 and on n−1 lines of degree n. Furthermore, L meets every line.

<u>Proof</u>. If some line has degree at least n+2, then Lemma 2.5 shows that $\mathbf{L} = \mathbf{E}_2$. We may therefore assume that every line has degree at most n+1.

Consider first the case that q is contained in two lines L and H of degree n+1. Because H has degree n+1, every point $\neq$ q of L has degree at least n+1, hence L meets at least n^2+n-1 other lines. In view of $b \le n^2+n$, we conclude that every point $\neq q$ of L has degree $n+1$, so Lemma 4.1 shows that L can be embedded into a projective plane of order n.

Now suppose that q lies on at most one line of degree $n+1$. Then $v \ge n^2-n+2$ implies that $v = n^2-n+2$ and that q lies on a unique line L of degree $n+1$ and on $n-1$ lines $N_2,\ldots,N_n$ of degree n.

Assume by way of contradiction that there exists a line H missing L. We may assume that H meets N_2 and N_3. Set $p_j = H \cap N_j$, $j = 2,3$. Since H misses L, the points p_j have degree at least $k_L+1 = n+2$. Since the points $\neq q$, p_1, p_2 of $N_2 \cup N_3$ have degree at least $k_L = n+1$, it follows that there are at least n^2+n+1 lines which meet N_2 or N_3. But we have $b \le n^2+n$.

This contradiction shows that L meets every other line, which implies that every point outside of L has degree $n+1$. If $b = n^2+n$, then (3) occurs. We may therefore assume that $b \le n^2+n-1$.

The line N_2 meets n^2-1 other lines and the line N_j, $j \ge 3$, meets $n-1$ lines which miss N_2. Thus $b = n^2+n-1$, N_2 is parallel to $n-1$ lines, and N_j, $j \ge 3$, meets each of the lines parallel to N_2. Since L meets every other line, we conclude that the lines parallel to N_2 have degree $n-1$. The same holds for N_j for all $j = 2,\ldots,n$. Hence there are $(n-1)^2$ lines of degree $n-1$ each of which is parallel to exactly one of the lines N_j. Now we already know the line L, the lines N_j and $(n-1)^2$ lines of degree $n-1$. The remaining $2n-2$ lines have to meet all the lines passing through q and have therefore degree n.

If every point of L has degree at most $n+1$, then $n-1$ of the points of L have degree $n+1$ and the other two have degree n (since the number of lines is n^2+n-1). If we remove q and add a line of degree one containing the other point of degree n, then we obtain an $(n+1,1)$-design with n^2-n+1 points and n^2+n lines. As in the proof of Lemma 4.1, it follows that D and L can be embedded in this case.

Finally consider the case that there is a point p of degree $r > n+1$ on L. Every line $\neq L$ through p has degree n or $n-1$. Thus, if t denotes the number of lines of degree n through p, then

$$n^2-2n+1 = v-k_L = t(n-1)+(r-1-t)(n-2) = (r-1)(n-2)+t.$$

Since $r > n+1$, we conclude that $n = 3$, $r = 5$ and $t = 0$ (notice that we have $n \geq 3$, since there are lines of degree $n-1$). If we remove p and the 4 lines of degree $n-1 = 2$ which pass through p, then we obtain a linear space with $n^2-n+1 = 7$ points and lines. This is the projective plane of order 2, so **L** is the linear space $\mathbf{E}_1$.■

Before we can determine the linear spaces satisfying (3) of Lemma 4.2, we need two further lemmata. The proof of the first one is straightforward.

<u>4.3 Lemma</u>. Let **P** be a projective plane of some order $n \geq 5$. Suppose that M is a set of $n-1$ lines of **P** and that X is a set of points of **P** such that every line of M contains exactly two points of X and any two distinct lines of M intersect in a point of X. Then the lines of M are concurrent.■

<u>4.4 Lemma</u> (Totten, 1976e). Let C be the pseudo-complement of two lines in a projective plane of order n. If $n \geq 5$, then C is in fact the complement of two lines in a projective plane of order n.

<u>Proof</u>. By our hypothesis, C is an $(n+1,1)$-design with n^2-n points, $n-1$ lines of degree n and n^2 lines of degree $n-1$. This implies that each point is contained in a unique line of degree n and in n lines of degree $n-1$. It follows that every n-line intersects every line of degree $n-1$.

Let L be any line of degree $n-1$ and denote by L_1 any line parallel to L. Then L_1 has degree $n-1$, and each point of L_1 lies on a second line which is parallel to L. Thus, L_1 meets $n-1$ other lines which are parallel to L. If L_2 and L_3 are two of these $n-1$ lines, then Lemma 3.7 a) shows that L_2 and L_3 are parallel. It follows that L and the $n-1$ lines which miss L and meet L_1 form a set Π of n mutually parallel lines of degree $n-1$. Since C has n^2-n points, Π is a parallel class of C. We may therefore add an infinite point ∞ to each line of Π and obtain the linear space $\mathbf{L} = C_\infty \Pi$ with n^2-n+1 points. The point ∞ is contained in n lines of degree n. If we adjoin the line $\{\infty\}$, then we obtain an $(n+1,1)$-design D, which lies in a projective plane **P** of order n (Lemma 3.4 b)). By the construction of D, also C lies in **P**. The parameters of C imply almost immediately that C is the complement of two lines in in **P**.■

Remarks. 1.) Obviously Lemma 4.4 is also true for $n = 3$. For $n = 4$, however, it is not. It was shown by Totten that there is a unique exception, which is called the Shrikhande finite linear space (see Totten (1976) and Shrikhande (1969)).

2.) Lemma 4.4 is a special case of a theorem of Mullin and Vanstone (1976b). They embedded the pseudo-complement of any number e of concurrent lines in a projective plane of order n provided that n is large compared with e.

4.5 Lemma. If (3) of Lemma 4.2 occurs, then $\mathbf{L}$ can be embedded into a projective plane of order n.

Proof. By hypothesis, $\mathbf{L}$ has a point q, which is contained in a line L of degree n+1 and in lines $N_2,\dots,N_n$ of degree n. Furthermore, L meets every other line. This implies that every point outside L has degree n+1 and that

(1) $$\sum_{p\in L}(r_p-n) = b-1-k_L(n-1) = n.$$

In view of Lemma 4.1, it suffices to show that every point of $L-\{q\}$ has degree n+1.

Assume to the contrary that some point of $L-\{q\}$ has not degree n+1. Then (1) implies that L has a point $q'\neq q$ of degree n (because every point has degree at least n). Since q also has degree n, every line other than L through q' has degree at most n. Since the number of points is $v = n^2-n+2$, it follows that q' is contained in lines $N'_2,\dots,N'_n$ of degree n.

Set $\mathcal{S} = \{p \mid p \text{ is a point outside of } L\}$, and $\mathcal{M}_j = \{N_j\}\cup\{X \mid X \text{ is a line missing } N_j\}$, $j = 1,\dots,n-1$. Because N_j meets n^2-1 other lines, we have $|\mathcal{M}_j| = n+1$. For distinct indices j and k, we have $|\mathcal{M}_j\cap\mathcal{M}_k| = 1$, since N_j meets $|N_j| = n$ lines of $\mathcal{M}_k$. Furthermore, every point of $\mathcal{S}$ has degree n+1 and is therefore contained in a unique line of each of the sets $\mathcal{M}_j$.

Let $p_2,\dots,p_n$ be n−1 new symbols. We define an incidence structure L' in the following way. The points of L' are the elements of $\mathcal{S}\cup\{p_2,\dots,p_n\}$ and the lines of L' are the lines other than L of $\mathbf{L}$. A point of $\mathcal{S}$ is incident with a line of L', if it is in $\mathbf{L}$, and the point p_j is incident with the lines of $\mathcal{M}_j$. Thus, L' is obtained from $\mathbf{L}$, if we remove the line L and the n points $\neq q$ of L and let the lines of $\mathcal{M}_j$ intersect in the new point p_j, $j = 2,\dots,n$.

In view of the above properties, L' is a linear space with n^2-n points and n^2+n-1 lines. Furthermore, every point of L' has degree n+1, and the lines N_j have degree n in L'.

Consider a line X of L not containing q. If X meets the line N_j in L, then it does in L'. If X is parallel to N_j in L, then $X \in \mathcal{M}_j$ so that X and N_j meet in L' in the point p_j. It follows that every line not containing q in L has degree n-1 in L'. Consequently, L is the pseudo-complement of two lines in a projective plane of order n.

Assume by way of contradiction that $n \geq 5$. Then Lemma 4.4 shows that L' is the complement of two lines in a projective plane P of order n. Since the n-1 lines N'_j are mutually parallel lines of degree n-1 in L', Lemma 4.3 shows that they intersect in P in a point x. Since L' is the complement of two lines in P, there exists a unique other line X of L' passing in P through x. In L', X is parallel to each line N'_j. By the construction of L', X is also a line of L, and X is also in L parallel to all lines N'_j. But this is not possible, since the lines N'_j, $j = 1,\ldots,n-1$, meet in L in the point q' of degree n (no line can be parallel to all but one of the lines passing through the point q').

Consequently, $n \leq 4$. Consider the linear space L. Then (1) and the equation (B4) (see chapter 1) show that

$$(2) \qquad \sum_{N_j \neq X \in \mathcal{M}_j} k_X = v - |N_j| = n^2-2n+2$$

for all $j \in \{1,\ldots,n-1\}$. Since q has degree n, every line other than N_j of $\mathcal{M}_j$ has degree at most n-1. Since every line has at least two points and in view of $n \leq 4$, this implies that every line other than N_j of $\mathcal{M}_j$ has degree n-2 or n-1. Therefore, (2) and $|\mathcal{M}_j| = n+1$ yield that $\mathcal{M}_j$ contains two lines of degree n-1 and n-2 lines of degree n-2. In particular, $n-2 \geq 2$, i.e. n = 4.

Since each 3-line lies in a unique set $\mathcal{M}_j$, there are exactly 2(n-1) = 6 lines of degree 3. Similarly, there are ½(n-1)(n-2) = 3 lines of degree 2. Since b = 20, L has therefore ten 4-lines. Six of them contain q or q'.

Let D be the (5,1)-design induced by the lines other than L on the nine points outside of L. Let us call a line $\neq$ L of L a block, if we consider it as line of D. D has ten blocks of degree 3, six blocks of degree 2, and three blocks of degree 1. If a point p of D is contained in a block of degree 1, then the other four blocks through p must have degree 3, since p is connected to each of the 8 other points of D. Similarly, a point which lies not on a block of degree 1 is contained in three blocks of degree 3 and in two blocks of degree 2. Thus, D has exactly three points which are contained in four blocks of degree 3.

If p is a point of the linear space L which lies on the line L, then the lines $\neq L$ through p form a parallel class of D. In this way, we obtain $k_L = 5$ parallel classes of D. q and q' correspond in this way to $\pi := \{N_1,N_2,N_3\}$ and $\pi' := \{N'_1,N'_2,N'_3\}$. Since every line intersects L, each block of D lies in exactly one of these parallel classes. Now we shall determine the structure of D to show that this is not possible.

Let A be an affine plane of order 3 which is imposed on the points of D such that π and π' are also parallel classes of A. Denote by π_1 and π_2 the other two parallel classes of A. If p_1,p_2,p_3 are three points of A any two of which are not contained in a line of $\pi \cup \pi'$, then it is easily verified that $\{p_1,p_2,p_3\}$ is a line of A. Hence, if L_1,L_2,L_3,L_4 are the four blocks $\notin \pi \cup \pi'$ of D, then the L_j are also lines of A so that $L_j \in \pi_1 \cup \pi_2$. Now there are two possibilities. Each set π_j contains two of the blocks L_j or one of the sets π_j contains three of the blocks L_j. However, the first possibility can not occur, since D has exactly three points which are contained in four blocks of degree 3. W.l.o.g. we may therefore assume that $\pi_1 = \{L_1,L_2,L_3\}$ and $L_4 \in \pi_2$. If H is one of the two lines of $\pi_2-\{L_4\}$ of A and if p,p' are two points on H, then the block pp' of D can not have degree 3, so $pp' = \{p,p'\}$. It follows that D can be obtained from A, if we 'break up' the two lines $\neq L_4$ of π_2 each into three blocks of degree two and if we add a block of degree one to the three points of L_4. Now we know the structure of D and we see that it has no parallel class which contains the block L_4.

This final contradiction completes the proof of Lemma 4.5.∎

Now we study the linear space L with $v \geq n^2-n+2$ points and a point of degree n, if it has n^2+n+1 lines. The reader should notice that L can not be embedded into a projective plane of order n in this case, because every linear space with n^2+n+1 lines which is embedded in a projective plane of order n has constant point degree n+1. It is therefore not surprising that there are not many examples for such linear spaces. From the following lemma we shall obtain easily that L has n^2-n+2 points.

4.6 Lemma. Suppose L has at least n^2-n+2 points, a point q of degree n, and n^2+n+1 lines. Suppose furthermore that q lies on a line G of degree n+1 which has n points of degree n+1. Then **L** is the linear space $\mathbf{E}_4$. In particular, G is the only line of degree n+1 passing through q.

Proof. We shall prove this lemma in several steps. The first step will be used frequently below without mentioning it.

Step 1. There is a unique line H parallel to G. Every point of H has degree n+2, q has degree n, and every other point has degree n+1. Every line not containing q has degree at most n.

For: G intersects n^2+n-1 lines. In view of $b = n^2+n+1$, this shows that there is a unique line H parallel to G. Since the points of H are contained in exactly one line parallel to G, they have degree $k_G+1 = n+2$. The other points outside of G are not contained in any line parallel to G and have therefore degree $k_G = n+1$. Clearly, the degree of a line not containing q can not exceed the degree of q.

Step 2. If N_1 and N_2 are parallel lines of degree n which do not contain q, then either both lines meet H or both lines are parallel to H.

For: Assume by way of contradiction that N_1 and H are parallel while N_2 and H meet in a point h.
Let h' be any point other than h of H, and denote by L_1 and L_2 the two lines through h' which are parallel to N_2. Since h' is also contained in exactly two lines parallel to N_1 and because H is one of these two lines, N_1 meets one of the lines L_1 and L_2. But every point of N_1 has degree n+1 and is therefore contained in a unique line parallel to N_2, which is N_1 itself, a contradiction.

Step 3. If there is a line N of degree n which does not contain q and which intersects H, then **L** is the linear space $\mathbf{E}_4$.

For: Let h be the point of intersection of N and H, and denote by $\mathcal{M}$ the set of lines parallel to N. Since N intersects n^2+1 other lines, $\mathcal{M}$ contains n−1 lines. Furthermore, by (B4),

$$(+)\qquad n^2-2n+k_H \le (v-n-1)+(k_H-1) = \sum_{p\notin N}(r_p-k_N) = \sum_{X\in \mathcal{M}} k_X.$$

Every line of M has degree at most n. Hence, if c denotes the number of lines of degree n contained in M, then (+) yields

$$n^2-2n+k_H \le |\mathcal{M}|(n-1)+c = n^2-2n+1+c.$$

Consequently, $\mathcal{M}$ contains a line N' of degree n. By Step 2, N' and H intersect in a point h'. Denote by L the line other than N' which contains h' and which is parallel to N.

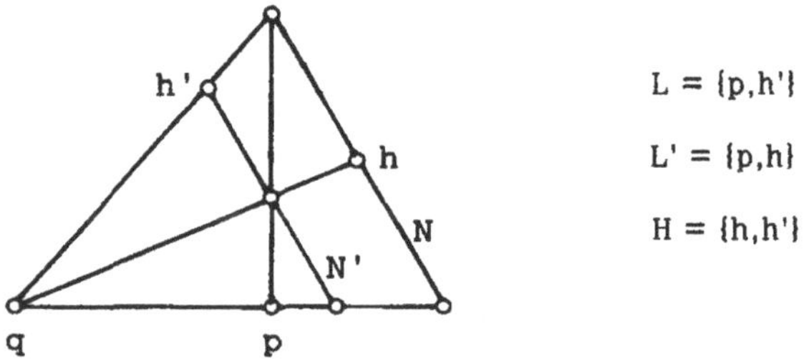

The linear space $\mathbf{E}_4$

If p is any point other than h' of L, then p has degree n+1 and is therefore contained in a line L' parallel to N'. Since L is the unique line through p which is parallel to N, the lines N and L' intersect. The point of intersection has degree at least n+2, since it is contained in two lines parallel to N'. Because h is the only point of degree $\neq$ n+1 of N, it follows that $h \in L'$, i.e. L' is the unique line other than N through h which misses N'. This shows that also p is uniquely determined, which implies that $p = G\cap L$, $L = \{p,h'\}$, and $k_L = 2$.
Now, (+) shows that

$$n^2-2n+k_H \le v-n-2+k_H = \sum_{X\in \mathcal{M}} k_X \le k_L+|\mathcal{M}-\{L\}|\cdot n = n^2-2n+2.$$

Consequently, $k_H = 2$, $v = n^2-n+2$, and every line other than L of $\mathcal{M}$ has degree n. By Step 2, each line of $\mathcal{M}$ intersects therefore H. Since the number of lines of $\mathcal{M}$ which intersect H is $2(k_H-1) = 2$, we conclude that n = 3 and v = 8.
The two lines other than G through q intersect N and N', hence they have degree at least three. In view of v = 8, they have degree three. Each point of G−{q} has

degree 4 and is therefore contained in one line of degree three and two lines of degree two. This shows that **L** is the linear space $\mathbf{E}_4$.

In view of Step 3, we may assume from now on that every line of degree n contains q or is parallel to H. We have to derive a contradiction.

Step 4. It exists n mutually disjoint sets $\mathcal{M}_1,\dots,\mathcal{M}_n$ with the following properties: $H \notin \mathcal{M}_j$, $|\mathcal{M}_j| = n$, and $\mathcal{M}_j$ is a partition of the points $\neq q$. $j = 1,\dots,n$.

For: Suppose we have already found sets $\mathcal{M}_1,\dots,\mathcal{M}_s$, $0 \le s \le n-1$, with the desired properties. Let $\mathcal{M}$ denote the set of lines $\neq H$ which do not contain q and which are not contained in $\mathcal{M}_j$, $j = 1,\dots,s$. Then $\mathcal{M}$ has n(n−s) lines each of which has degree at most n. Since each of the $v-1 > n(n-1)$ points other than q is contained in n−s of the lines of $\mathcal{M}$, this shows that $\mathcal{M}$ contains a line N of degree n. By the definition of $\mathcal{M}$, N does not contain q. Thus, N is parallel to H. Let $\mathcal{M}_{s+1}$ denote the set consisting of N and the n−1 lines other than H which are parallel to N. Since the point q has degree n, it is not contained in any of the lines of $\mathcal{M}_{s+1}$. In view of Step 1, every other point is contained in a unique line of $\mathcal{M}_{s+1}$. Since N intersects each of the n lines of $\mathcal{M}_j$, $\mathcal{M}_{s+1}$ is disjoint to $\mathcal{M}_j$, $j = 1,\dots,s$. This proves Step 4.

Now, we extend **L** to a linear space **L'** in the following way. We let the lines of $\mathcal{M}_j$ intersect in a new point q_j, $j = 1,\dots,n$, and we add the new line $\{q,q_1,\dots,q_n\}$. It follows from Step 4 that **L'** is in fact a linear space. It has $v' := v+n \ge n^2+2$ points and $b' := b+1 = n^2+n+2$ lines. Furthermore every point of H has degree n+2 and every other point has degree n+1 in **L'**. Obviously, every line of **L'** has degree at most n+1. Fix a point p of H, which has degree $r := n+2$.

Assume that p lies on a line X of degree n+1 in **L'**. Then $X \neq H$, since H has in **L** and **L'** the same degree which is at most $r_q = n$. Hence, every point $\neq p$ of X has degree n+1 in **L'** so that X meets n^2+n+1 other lines. Since $b' = n^2+n+2$, it follows that X meets every line. However each point of $H-\{p\}$ has degree n+2 and lies on a line missing X, a contradiction.

Consequently every line through p has degree at most n. Hence, if $L_1,\dots,L_r$, $r := n+2$, are the lines through p and if we define the numbers S_j as in Lemma 2.2 (for the linear space **L'**), then $S_j < 1$ for all j. Lemma 2.2 implies therefore that $b' \ge v'+r-1 = v'+n+1$, which is a contradiction.■

<u>4.7 Theorem</u>. Suppose L is a linear space with at least n^2-n+2 points and at most n^2+n+1 lines for some integer $n \geq 2$. If L has a point of degree at most n, then one of the following cases occurs:

(1) L is a near-pencil.

(2) L can be embedded into a projective plane of order n.

(3) L is one of the linear spaces $\mathbf{E}_1$, $\mathbf{E}_2$, $\mathbf{E}_3$ and $\mathbf{E}_4$; in particular $n = 3$ and $v = n^2-n+2 = 8$.

(4) $v = n^2-n+2$ and $b = n^2+n+1$. L has a point q of degree n which lies on one line L of degree n+1 and on n−1 lines $N_2,\dots,N_n$ of degree n. The n points $\neq q$ of L do not all have degree n+1.

<u>Proof</u>. If there is a line of degree at least n+2, then Lemma 2.5 shows that L is a near-pencil or $\mathbf{E}_2$ or $\mathbf{E}_3$. If $b \leq n^2+n$, then Theorem 4.7 follows from Lemma 4.2 and Lemma 4.5. Assume from now on that $b = n^2+n+1$ and that every line has degree at most n+1.

Let q be a point which has degree at most n. In view of $v \geq n^2-n+2$, it follows that q has degree n and that q lies on a line L of degree n+1.

Assume by way of contradiction that q is contained in a second line L' of degree n+1. Then every point $\neq q$ of L' and L has degree at least n+1. Lemma 4.6 shows therefore that there are points $p \in L$ and $p' \in L'$ of degree at least n+2. But now L meets at least n^2+n other lines and L is parallel to a line passing through p'. Thus there are at least n^2+n+2 lines, a contradiction.

Consequently, L is the only line of degree n+1 through q, so $v \geq n^2-n+2$ implies that $v = n^2-n+2$ and that the n−1 lines other than L through q have degree n. If every point $\neq q$ of L has degree n+1, then Lemma 4.6 shows that $L = \mathbf{E}_4$, and if not, then (4) is fulfilled.■

<u>4.8 Corollary</u>. In a non-degenerate linear space with n^2+n+1 lines and at least n^2-n+3 points, every point has degree at least n+1.■

Case (4) of Theorem 4.7 is particularly difficult. We shall be occupied with it in chapter 12.

5. Linear spaces with few lines

In Theorem 2.6, we proved $b \geq B(v)$ for every non-degenerate linear space. If n is the order of a projective plane P, then examples of linear spaces with v points and $b = B(v)$ lines can be obtained easily from P for each value of v with $n^2-n+2 \leq v \leq n^2+n+1$. But do we obtain all examples in this way? Almost yes: Every non-degenerate linear space other than E_1 with $n^2-n+2 \leq v \leq n^2+n+1$ and $b = B(v)$ lies in a projective plane of order n. This will be shown in this chapter. Each of the cases $v = n^2-n+2 \neq 4$, $n^2+2 \leq v$, and $n^2-n+3 \leq v \leq n^2+1$ or $v = 4$ will be studied separately. In this chapter, we shall furthermore classify the linear spaces with $v = n^2-n+2$ and $b = n^2+n$ so that the problem of the determination of linear spaces with $v \geq n^2-n+2$ and $b \leq n^2+n+1$ is reduced to the case $b = n^2+n+1$.

5.1 Lemma (De Witte, 1977b). Let L be a non-degenerate linear space. If L has at least n^2 points and at most n^2+n+1 lines for some integer $n \geq 2$, then it can be embedded into a projective plane of order n.

Proof. In view of Lemma 2.5, every line has degree at most n+1, and in view of Theorem 4.7 we may assume w.l.o.g. that every point has degree at least n+1. We consider two cases.

Case 1. There is a line L of degree n+1.
If every point of L has degree n+1, then L meets n^2+n other lines. Since $b \leq n^2+n+1$, we conclude that $b = n^2+n+1$, that every point of L has degree n+1, and that L meets every other line. This last fact implies that every point outside of L has degree n+1. Hence, L is an (n+1,1)-design. Lemma 3.5a) shows that L can be embedded into a projective plane of order n.

Case 2. Every line has degree at most n.
Then every point p has degree at least n+1 with equality if and only if $v = n^2$ and every line through p has degree n. In view of

$$v(n+2) > (n^2+n+1)n \geq \sum_{L \in \mathcal{L}} k_L = \sum_{p \in \mathcal{P}} r_p,$$

there is a point q of degree n+1. Hence, $v = n^2$ and the lines $N_1, \ldots, N_{n+1}$ through q have degree n.

Assume by way of contradiction that some point p has degree at least n+2. W.l.o.g. $p \in N_1$. Then N_1 meets at least n^2+1 other lines, so N_1 is parallel to at most n−1 lines. Since each of the n−1 points $x \neq q$ of N_j, $j \geq 2$, lies on $r_x-n \geq 1$ lines parallel to N_1, we conclude that every point outside N_1 has degree n+1. Hence, every line contains a point of degree n+1 and has therefore degree n. This implies $v-1 = r_p(n-1)$ for every point p, a contradiction, since q has degree n+2.

Consequently, every point has degree n+1. This yields that every line has degree n so that L is an affine plane of order n.■

5.2 Lemma (Erdös, Mullin, Sós and Stinson, 1983). If L is a linear space with n^2-n+2 points and n^2+n-1 lines, then $L = E_1$ or L can be embedded into a projective plane of order n.

Proof. In view of Theorem 4.7, it suffices to show that some point has degree at most n. Assume to the contrary that every point has degree at least n+1. Then $1+k_L \cdot n \leq b < n^2+n+1$ and hence $k_L \leq n$ for every line L. Since

$$\sum_{L \in \mathcal{L}} k_L = \sum_{p \in \mathcal{P}} r_p \geq v(n+1) = n^3+n+2 > b(n-1)$$

it follows that there exists a line N of degree n. N meets at least n^2 other lines. Because every point outside of N lies on at least one line parallel to N, we have $v-k_N \leq nc$ for the number c of lines missing N. Hence $b \geq |\{N\}| + n^2+c > n^2+n-1$, which is a contradiction.■

Remark. Suppose L is a linear space $\neq E_1$ with n^2-n+2 points and n^2+n-1 lines. By Lemma 5.2, L can be embedded into a projective plane P of order n. Let H and L be the two lines of P which are not lines of L. Since any two points of L lie on a line of L, at most one of the points of H and one of the points of L is a point of L. In view of $v = n^2+n+1-(2n-1)$, this implies that there are points $p \in H$ and $q \in L$ with $p,q \neq H \cap L$ such that L is the complement of $H \cup L - \{p,q\}$ in P. The structure of L is clear now.

Before we can determine the linear spaces with $n^2-n+3 \leq v \leq n^2+1$ and $b = B(v) = n^2+n$, we need the following lemma.

<u>5.3 Lemma</u>. Let L be a linear space with $n^2-n+1 \le v \le b = n^2+n$ for some integer n. Suppose that every point has degree at least n+1 and that some point has degree at least n+2. Then L is the closed pseudo-complement of a Baer-subplane in an affine plane of order n. In particular, $v = n^2-n+1$.

<u>Proof</u>. For each n-line N, we set $\mathcal{M}_N = \{N\} \cup \{X \in \mathcal{L} \mid X \cap N = \emptyset\}$, and call such a set a clique. Let s denote the number of cliques. We shall prove Lemma 5.3 in several steps.

Step 1. Every line has degree at most n.
For: Since every point has degree at least n+1, this follows from $b < n^2+n+1$.

Step 2. Every clique contains exactly n lines, and any two distinct cliques are disjoint. In particular, every point of an n-line has degree n+1.
For: Let m denote the number of lines of a clique $\mathcal{M}_N$. Obviously,

$$m = b - \sum_{p \in N} (r_p - 1) \le b - n^2 = n$$

with equality if and only if every point of N has degree n+1. Since every point of L is contained in at least one of the lines of $\mathcal{M}_N$, we have also $mn \ge v > n^2-n$. Thus m = n and every point of N has degree n+1.
Let $G \neq N$ be any line of degree n. Then also every point of G has degree n+1. Consequently, if $N \cap G = \emptyset$, then $\mathcal{M}_N = \mathcal{M}_G$. If G and N intersect, then $\mathcal{M}_N \cap \mathcal{M}_G = \emptyset$, since G meets each of the n lines of $\mathcal{M}_N$.

Step 3. If n+1+f is the degree of a point p, then $s(f+1) \le n+1+f$.
For: Each clique contains f+1 lines passing through p (note that p does not lie on an n-line, if $f \ge 1$). Since distinct cliques are disjoint, the assertion follows.

Step 4. Suppose a line N of degree n is parallel to lines L_1 and L_2 which intersect in a point q. If L_j has degree $n+1-d_j$, j = 1,2, then $n \le d_1 d_2$.
For: This follows from the Parallel-Lemma (see section 1).

Step 5. (a) If q is a point with $r_q > n+1$, then q has degree $n+\sqrt{n}$. Every line passing through q has degree $n+1-\sqrt{n}$ and is contained in a unique clique.
(b) $v = n^2-n+1$ and $s = 1+\sqrt{n}$. In particular, n is a perfect square.

(c) Of the lines passing through a point of degree $n+1$, $\sqrt{n}$ have degree n, $n-\sqrt{n}$ have degree $n-1$, and the last one has degree $n+1-\sqrt{n}$.

For: We prove all parts together. Set $d_L = n+1-k_L$ for every line L.

By our assumptions, there exists a point q which has degree at least $n+2$. Set $r_q = n+1+f$, denote by L a line of maximal degree through q, and set $d = d_L$. Step 2 shows that $d \geq 2$. From

$$n^2-n \leq v-1 \leq r_q(n-d) = n^2+n-n(d-f)-d(f+1) < n^2+n-n(d-f),$$

we obtain $d \leq f+1$. Set $\mathcal{M} = \{X \in \mathcal{L} \mid X \cap L = \emptyset\}$. We have

$$|\mathcal{M}| = b-1-\sum_{p \in L}(r_p-1) \leq b-1-(k_L-1)n-(r_q-1) = dn-1-f \leq d(n-1)$$

and

$$\sum_{X \in \mathcal{M}} k_X = \sum_{p \notin L}(r_p-k_L) \geq (v-k_L)d > (n-1)^2d,$$

Together it follows that $\mathcal{M}$ contains a line N of degree n. Let p' be a point of N. Then p' has degree n (Step 2). In view of $v-1 \geq r_{p'}(n-2)+2$, p' lies on a second line G of degree n.

Let H be any line other than L of $\mathcal{M}_N$ passing through q, and put $k_H = n+1-t$. By Step 4, $n \leq dt$.

In view of $H \in \mathcal{M}_N$, Step 2 shows that G and H meet in a point p of degree $n+1$. If c denotes the number of lines of degree n passing through p, then

$$n^2-n+1 \leq v \leq k_H+c(n-1)+(r_p-1-c)(n-2) = n^2-n+1+c-t.$$

Thus, $c \geq t$. Each of the c lines of degree n through p determines a clique. All these cliques do not contain H and are therefore different from $\mathcal{M}_N$. Hence, there are at least $c+1$ cliques, i.e. $s \geq c+1$. Now, $s(f+1) \leq r_q = n+1+f$ (Step 3), $dt \geq n$, and $d \leq f+1$ show

$$\frac{n}{d} \leq t \leq c \leq s-1 \leq \frac{n}{f+1} \leq \frac{n}{d}.$$

Thus, we get equalities for the above inequalities. This shows that

(1) $v = n^2-n+1$, $c = t = s-1$, $d = f+1$, $n = dt$, and

(2) $r_q = s(f+1)$, i.e. every line passing through q is contained in a clique (see the proof of Step 3).

Since H was an arbitrary line other than L of M_N through q, each of the f lines other than L of M_N through q has degree n+1−t. Let $M_1 = M_N, M_2, \ldots, M_w$, $1 \leq w \leq s$, denote the cliques which contain a line of degree n+1−d through q. As for M_N, we can show that

(3) of the f+1 lines of M_j which pass through q, one has degree n+1−d and the others have degree n+1−t (j = 1,...,w).

Since distinct cliques are disjoint and because of $d \geq 2$, we conclude

$$\sum_{q \in X} d_x \geq w(d+ft) = w(d+n-\frac{n}{d}) > w\frac{n}{d}.$$

In view of

$$\sum_{q \in X} d_x = r_q n - \sum_{q \in X}(k_x-1) = r_q n - (v-1) = n(d+1),$$

we obtain $w < d^2+d$.

Let e be the number of lines of degree n+1−d passing through q. Since n+1−d is the maximal degree of the lines passing through q, we have

$$n^2-n = v-1 = \sum_{q \in X}(k_x-1) \leq e(n-d)+(r_q-e)(n-d-1)$$
$$= r_q(n-d-1)+e = (n+d)(n-d-1)+e = n^2-n-d^2-d+e,$$

i.e. $e \geq d^2+d > w$.

Since, by (2), each of the e lines of degree n+1−d through q is contained in one of the w cliques M_j, we conclude that one of the cliques M_j contains at least two lines of degree n+1−d. (3) shows therefore that $d = t$. We conclude $n = d^2$ and $r_q = n+1+f = d^2+d \leq e$. Hence $r_q = e$ so that every line passing through q has degree n+1−d.

It remains to prove part (c). Let x be any point of degree n+1. The line xq has degree $n-\sqrt{n}$ and is contained in a clique. Each line of degree n through x determines another clique. Since the number of cliques is $s = \sqrt{n}+1$, it follows that x is contained in at most $\sqrt{n}$ lines of degree n. In view of

$$v = n^2-n+1 = k_{xq}+\sqrt{n}(n-1)+(r_x-1-\sqrt{n})(n-2),$$

it follows that x lies on the line xq, on $\sqrt{n}$ lines of degree n, and on $n-\sqrt{n}$ lines of degree n-1. This proves (c).

Step 6. There is a unique point of degree $n+\sqrt{n}$.
For: Assume to the contrary, there are two points q_1 and q_2 of degree $n+\sqrt{n}$. Since there are cliques, there exists a line N of degree n. Choose a point p of N with $pq_1 \neq pq_2$. Since p has degree n+1, each clique contains a unique line passing through p. By Step 5 (a), the lines pq_1 and pq_2 are contained in a clique. By Step 5 (c), p is also contained in $\sqrt{n}$ lines of degree n, each of which determines a clique. Thus, at least $\sqrt{n}+2$ of the lines through p are contained in a clique. This contradicts Step 5 (b).

Step 5 and Step 6 show that there is a unique point q of degree $n+\sqrt{n}$ and that every other point has degree n+1. Every line passing through q has degree $n+1-\sqrt{n}$ and every other line has degree n-1 or n. Removing q we obtain therefore an (n+1,1)-design D, which is the pseudo-complement of a Baer-subplane in an affine plane of order n. If D is in fact a complement, then **L** is the closed complement of a Baer-subplane in an affine plane. Since we do not know this (but see Lemma 5.9), we can just state that **L** is the closed pseudo-complement of a Baer-subplane in an affine plane of order n.■

<u>5.4 Lemma</u>. Every non-degenerate linear space **L** with $v \geq n^2-n+3$ points and n^2+n lines for some integer $n \geq 2$ can be embedded into a projective plane of order n.

<u>Proof</u>. If every point has degree n+1, this follows from Lemma 3.5, and if some point has degree at most n, it follows from Theorem 4.7. Lemma 5.3 shows that no other case occurs.■

Now we have the following theorem, which is the full generalization of the Theorem of de Bruijn and Erdös.

<u>5.5 Theorem</u> (Metsch, 1991a). Let **L** be a non-degenerate linear space and denote by n the unique integer with $n^2-n+2 \leq v \leq n^2+n+1$. Then $b \geq B(v)$ and equality implies that $\mathbf{L} = \mathbf{E}_1$ or that **L** can be embedded into a projective plane of order n.■

What about linear spaces with $b = B(v)+1$? If $n^2+2 \leq v$, then the Theorem of Totten (see chapter 8) shows that L is Lin's Cross. For $n^2-n+3 \leq v \leq n^2+1$, the problem is much more difficult. We shall determine only the linear spaces with $n^2-\frac{1}{2}n+6 < v \leq n^2+1$ and $b = B(v)+1 = n^2+n+1$. However, they can not all be embedded into a projective plane of order n. This will be shown in the chapter 7. But we have already determined the linear space satisfying $v = n^2-n+2 \neq 4$ and $b = B(v)+1 = n^2+n$:

5.6 Lemma. If L is a non-degenerate linear space with $v = n^2-n+2$ points and $b = B(v)+1 = n^2+n$ lines for some integer $n \geq 2$, then $L = \mathbf{E}_2$ or L can be embedded into a projective plane of order n.

Proof. As in the proof of Lemma 5.4 this follows from Lemma 5.3, Theorem 4.7, and Lemma 3.5.■

5.7 Corollary. If L is a linear space with at least n^2-n+2 points and at most n^2+n lines, then one of the following cases occurs:

(1) L is a near-pencil.

(2) L can be embedded into a projective plane of order n.

(3) $L = \mathbf{E}_1$ or $L = \mathbf{E}_2$.

Proof. If $v = n^2-n+2$ and $b = n^2+n$, this follows from Lemma 5.6. In every other case we have $b = B(v)$ so that Lemma 5.7 follows from Lemma 5.4.■

Corollary 5.7 completes the work we wanted to do in this section. However, it does not summarize all of our results. In Lemma 5.3, we obtained a characterization for the closed pseudo-complement of a Baer-subplane in an affine plane of order n. They are in fact complements, if $n \neq 4,9$. In order to prove this, we first have to embed the pseudo-complement of a Baer-subplane into a projective plane.

5.8 Theorem. Every pseudo-complement L of a Baer-subplane in a projective plane of order $n = m^2$ is in fact the complement of a Baer-subplane.

Proof. By the assumption, L has n^2-m points of degree $n+1$, n^2-m lines of degree n, which we call *long* lines, and $n+m+1$ lines of degree $n-m$, which we call *short*. It follows that each point is contained in a unique short line and in n long lines.

For each long line N, the set π_N consisting of N and the lines parallel to N is a parallel-class with $|\pi_N| = b-k_N n = n+1$. We call it a *clique*.

Since a long line meets n short lines, a clique has m+1 short an n−m long lines. If N and G are parallel long lines, then we have $\pi_N = \pi_G$. If N and G are intersecting long lines, then G intersects n of the lines of π_N so that π_N and π_G have a unique line in common; in this case, the line S of $\pi_N \cap \pi_G$ is short (if S were long, then we would obtain $\pi_N = \pi_S = \pi_G$). Hence, every two distinct cliques have a unique line in common, which is short.

Let $\pi_1,\ldots,\pi_s$ be the cliques. Since there are n^2-m long lines and because every clique has n−m long lines, we have $s(n-m) = n^2-m$, i.e. $s = n+m+1$. Let the lines of π_j intersect in an infinite point p_j, $j = 1,\ldots,s$. Since any two distinct cliques have a unique line in common, the obtained structure P is a linear space. It has $v+s = n^2+n+1$ points and lines and is therefore a projective plane. Consider the linear space B which is induced by P on the n+m+1 infinite points of L. Because, every long line of L has a unique infinite point, the lines of B are the n+m+1 short lines of L. Consequently, B is a projective plane of order m and L is the complement of this Baer-subplane in P.■

Theorem 5.8 is a special case of a theorem in (Metsch, 1988c), which says that the pseudo-complement of any number e of mutually disjoint Baer-subplanes can be embedded into a projective plane of order n. The case e = 2 was handled before by Batten and Sane (1985) and Drake (1985).

The affine case is much more difficult.

<u>5.9 Lemma</u>. Suppose L is the pseudo-complement of a Baer-subplane in an affine plane of order $n = m^2$. If $n \neq 4,9$, then L is in fact the complement of a Baer-subplane in an affine plane of order n.

<u>Proof</u>. By hypothesis, L has n^2-n points of degree n+1, (m+1)(n−m) lines of degree n, (n−m)n lines of degree n−1, and n+m lines of degree n−m.

First we sppose that L can be embedded into a projective plane P of order n and we show that L is then the complement of a Baer-subplane in an affine plane. The projective plane P has a unique line G which is not a line of L. Since L has constant point degree n+1, no point of G is a point of L. Let *q* be the set of the $n = m^2$ points of P which are not points of L and which do not lie on G. If L of is a line of degree n−d of L, then L has d+1 points in P−L. One of these points is $G \cap L$ and the other d points lie in *q*. This implies that any two points of *q* are connected by a line of degree n−m, and that any line of degree n−m has m points in *q*. Hence, the $n+m+1 = m^2+m+1$ lines of degree n−m induce an affine plane B of order m on the m^2 points of *q*. This shows that L is the complement of the Baer-subplane B in the affine plane A = P−G.

Now we show that L can be embedded. We distinguish two cases.

Case 1. There is a line N of degree n which is parallel to a line L of degree n−1. The set Π consisting of N and the lines parallel to N is a parallel-class with $|\Pi| = b-k_N n = n$ and $L \in \Pi$. Because every point outside of L is contained in two lines parallel to L, the set Π' consisting of L and the lines parallel to L which are not in Π is also a parallel-class. Since L is parallel to $b-k_L n = 2n-1$ lines and in view of $|\Pi| = n$, we have $|\Pi'| = n+1$. If we let the lines of Π' intersect in an infinite point, then we obtain a linear space L' with n^2-n+1 points, n^2+n lines and constant point degree n+1. By Lemma 3.6, L' can be embedded into a projective plane P of order n. Consequently also L lies in P.

Case 2. Every n-line meets every line of degree n−1.
Call the lines of degree n−m *short* and the lines of degree n−1 *long* (don't worry that the n-lines are still longer). Since every line parallel to a long line is short or long, Lemma 3.1 shows that a long line is parallel to m+1 short and 2n−m−2 long lines. Call a set *M* of at least n−m+1 mutually parallel long lines a *clique*. We shall prove certain properties of cliques now.

(1) Any two parallel long lines H and L are contained together in a clique.
For: Since H meets n−1 lines which are parallel to L and because L is parallel to only $\alpha := m+1$ short lines, H meets a long line G which is parallel to L. Let *M'* be the set of long lines which meet G and miss L. Then $|M'| \geq k_G-\alpha = n-2-m$ and $H \in M'$. By Lemma 3.7 a), any two lines of *M'* are parallel. Hence $M := M' \cup \{L\}$ is a clique.

(2) Any two distinct maximal cliques M_1 and M_2 have at most one line in common.

For: Since M_1 and M_2 are distinct maximal cliques, there are lines $L_j \in M_j$ with $L_1 \cap L_2 \neq \emptyset$. Since the number of lines parallel to L_1 and L_2 is three, and because every line in $M_1 \cap M_2$ is parallel to L_1 and L_2, we have $|M_1 \cap M_2| \leq 3$.
Assume by way of contradiction that $M_1 \cap M_2$ contains two lines H_1 and H_2. Then every line of $M_1 \cup M_2 - \{H_1, H_2\}$ is parallel to H_1 and H_2. Since the number of lines parallel to H_1 and H_2 is $n-1$, we obtain $2(n-1-m) \leq |M_1| + |M_2| = |M_1 \cap M_2| + |M_1 \cup M_2| \leq 3+(n+1)$ and thus $n \leq 2m+6$, which contradicts $n \geq 16$.

(3) Every long line is contained in exactly two maximal cliques M_1 and M_2. We have $M_1 \cap M_2 = \{L\}$, $|M_1| + |M_2| = 2n-m$ and $|M_j| \leq n$.
For: By (1), each of the $2n-2-m$ long lines parallel to a given long line L is contained in a maximal clique which also contains L. By (2), any two maximal cliques containing L have only the line L in common. Thus, if $M_1, \ldots, M_t$ are the maximal cliques containing L, then $2n-2-m = \sum |M_j - \{L\}|$. Since every clique has at least $n-m-1$ lines and in view of $3(n-m-2) > 2n-2-m$, we have $t \leq 2$, and because every clique has at most n lines (this follows from $v = n^2-n$), we have $t \geq 2$. This proves (3).

If M_1 and M_2 are the two maximal cliques containing a long line L, then we call $\{|M_1|, |M_2|\}$ the *type* of L.

(4) Not every long line has type $\{c,c\}$ where $2c = 2n-m$.
For: Assume to the contrary that every long line has type $\{c,c\}$. Then every maximal clique has exactly c elements. Since each of the $(n-m)n$ long lines is contained in two maximal cliques, it follows that c divides $2(n-m)n$. But then $2n-\sqrt{n} = 2n-m = 2c$ divides $4(n-m)n = (2n-m)^2-n$, which is not possible.

(5) Parallel long lines have the same type.
For: Since any two parallel long lines are contained in a common clique, this follows from (3).

(6) Every long line has type $\{n, n-m\}$.
For: By (4), there is a long line L of type $\{e,f\}$ with $e \neq f$. Let M_1 and M_2 be the maximal cliques containing L and assume w.l.o.g. $|M_1| = e$ and $|M_2| = f$. For each line X of M_1 let M_X be the second maximal clique containing X. By (4), X and

every line of M_X has type $\{e,f\}$. Since $e \neq f$, it follows that the e cliques M_X, $X \in M_1$, are mutually disjoint. Together they contain ef lines. In view of $e+f = 2n-m$ and $e,f \leq n$, we have $ef \geq n(n-m)$ with equality if and only if $\{e,f\} = \{n,n-m\}$. Because the number of long lines is $n(n-m)$, it follows that $\{e,f\} = \{n,n-m\}$ and that every long line is contained in exactly one of the cliques M_X, $X \in M_1$. Hence, every long line has type $\{e,f\}$.

(7) L can be embedded into a projective plane of order n.

For: By (6), each long line is contained in a unique clique M with $|M| = n$. In view of $v = n^2-n$, these cliques are parallel classes of L. Since there are $n(n-m)$ long lines, there are $n-m$ such parallel-classes. By (3), any two of these parallel classes are disjoint. We let the lines of each of these parallel-classes intersect in an infinite point and add an infinite line H which contains the $n-m$ infinite points. The obtained structure D has n^2-m points and n^2+n+1 lines. The long lines of L have degree n in D and H has degree $n-m$. Hence, D has $n+m+1$ lines of degree $n-m$ and n^2-m lines of degree n. By Lemma 5.8, D is the complement of a Baer-subplane B in a projective plane P of order n, so L is embedded in P. (In this case we see directly that $L = D-H$ is the complement of a Baer-subplane in the affine plane $A := P-H$.)■

<u>Remark</u>. Let L_∞ be a line of a projective plane P of order $n = m^2$, and define the affine plane $A = P-L_\infty$. Suppose B' is an (affine) Baer-subplane of A, and let L be the complement of B' in A. Consider the condition

(C_1) every n-line of L meets every line of degree $n-1$ of L.

If (C_1) holds, then we saw in the preceding proof that B' can be extended to a Baer-subplane B of P and that the infinite points of B', i.e. the points of B−B', lie on the line L_∞. (We remark that an affine Baer-subplane of A can always be extended to a projective Baer-subplane of P provided that P is a desarguesian projective plane of order at least 11. More generally, Rigby (1965) showed that for any affine subplane D of order $d \geq 4$ of a desarguesian projective plane P, the projective closure of D is embedded in P, i.e. P has a projective subplane of order d in which D is embedded. This result does not hold for $d = 3$: PG(2,4) contains an affine subplane of order 3, induced on the points of an hermitian unital, but PG(2,4) has no projective subplane of order 3). Condition (C_1) implies that

(C_2) the lines of the Baer-subplane B' are mutually parallel in L, hence they form a partition of the points of L.

Condition (C_2) says that we can extend L to the closed complement L_c of B' in A (we let the lines of B' intersect in a new point). Since the infinite points of B' are also infinite points of A, it follows that L_c is embedded in the closed complement of B in P.

Now suppose that L_c is the closed complement of B' in A. By definition, this implies that C_2 is fulfilled, and it is easy to see that also (C_1) is then fulfilled. Hence, we can extend L_c to a closed complement of a Baer-subplane in a projective plane of order n.

<u>5.10 Corollary</u>. Every closed pseudo-complement L of a Baer-subplane in an affine plane of order $n = m^2 \geq 16$ is in fact the closed complement of a Baer-subplane in an affine plane of order n. It can be embedded into the closed complement of a Baer-subplane in a projective plane of order n.

<u>Proof</u>. By hypotheses L has a point q of degree n+m while every other point has degree n+1. Furthermore L has n+m lines of degree n−m+1, n(n−m) lines of degree n−1 and (m+1)(n−m) lines of degree n. In view of $v = n^2-n+1=1+r_q(n-m)$, the lines of degree n−m+1 are the lines through q. The linear space D := L−q is therefore pseudo-complement of a Baer-subplane in an affine plane of order n. Theorem 5.10 shows that D is in fact a complement of a Baer-subplane in an affine plane of order n, and the preceding remark shows that L can be extended to the closed complement of a Baer-subplane in a projective plane of order n.■

<u>5.11 Corollary</u>. Suppose L is a linear space with n^2-n+1 points and n^2+n lines and every point of L has degree at least n+1. If $n \neq 4,9$ and if some point has degree at least n+2, then L is the closed complement of Baer-subplane in an affine plane of order n.

<u>Proof</u>. This follows from Lemma 5.3 and Corollary 5.10.■

6. Embedding (n+1,1)-designs into projective planes

In chapter 2 we proved that every (n+1,1)-design with $n^2-n+1 \le v \le b \le n^2+n$ can be embedded into a projective plane of order n. In this section, we consider the case $b = n^2+n+1$. This is the much more difficult case, and there are a lot of published results concerning it. The first one is a theorem of Vanstone (1973), (see Lemma 3.5a)), which shows that $v \ge n^2$ implies embeddability. The bound for v has been improved to $v \ge n^2-1 \neq 35$ (Bose and Shrikhande, 1973), $v \ge n^2-1 = 35$ (de Witte, 1977a), $v \ge n^2-2$ and $4v > 4n^2-3+\sqrt{(8n-7)}$ (McCarthy and Vanstone, 1977), $v > n^2+6-2\sqrt{(n+3)}$ (Dow, 1983, see also Dow, 1982), and $v > n^2-n/6$ (Metsch, 1988a). We are going to prove that it suffices to demand $v > n^2-\frac{1}{2}n+6$ and we shall show that there is a still better bound for large n (Corollary 6.14 and Theorem 6.15). Since $n^2-\frac{1}{2}n+6 \ge n^2-1$ for small values of n, we shall also give a proof of the Theorem of McCarthy and Vanstone, but only in the case $n \neq 6$. The case $n = 6$ will be handled in section 13.

The basic techniques for the proofs will be developed in the following proposition.

<u>6.1 Proposition</u>. Let L be an (n+1,1)-design with n^2+n+1 lines for some integer $n \ge 7$ and denote its number of points by $n^2+n+1-s$. Let $e \ge 2$ be any integer with $(e+1)^2 \le n+2$ and denote by α the largest integer satisfying $\alpha(e-1) \le 2(s-2-n)$. Call the lines of degree at most $n-e$ *short* and the other lines *long*, and suppose that the following conditions are satisfied:

(1) On each long line there is a point which is not contained in any short line.

(2) There is no line of degree n.

(3) There are at most n short lines.

(4) $2s \le 3n+2$.

(5) $n \ge e^2-2e+5+2\alpha$, $n \ge e^2+4+\alpha$, and $n \ge 3e+1+2\alpha$.

(6) If $e \ge 3$ then $2(e+1)n-5e+32 \ge s(e+3)$.

Then L can be embedded into a projective plane of order n.

We shall prove this proposition in Lemmata 6.2 to 6.10 in which L is always assumed to be a linear space satisfying the conditions of Proposition 6.1. We note that the integer α is not negative. This follows from (2) and Lemma 3.3.

A set of mutually parallel long lines is called a *clique*. Call a maximal clique $\mathcal{M}$ a *good* clique if $\mathcal{M}$ contains a line of degree $n-1$ and if $|\mathcal{M}| \geq n+1-\alpha$. We denote the degree of a line L by $k_L := n+1-d_L$.

<u>6.2 Lemma</u>. Suppose that L is a line of degree $n-1$.

a) There are at most α short lines parallel to L.

b) If H is parallel to L, then H meets a line of degree $n-1$ which is parallel to L.

<u>Proof</u>. a) The number of lines parallel to L is 2n. Let a denote the number of short lines parallel to L. Since each line parallel to L has degree at most $n-1$, Lemma 3.1 c) yields $2(n^2+2-s) = 2(v-k_L) \leq (2n-a)(n-1)+a(n-e) = 2n(n-1)-a(e-1)$. Hence $a(e-1) \leq 2(s-2-n)$, so the definition of α implies $a \leq \alpha$.

b) Let a be the number of lines $\neq$ H which miss L and have degree at most $n-2$. As before Lemma 3.1 c) shows that $2(n^2+2-s) \leq k_H+a(n-2)+(2n-1-a)(n-1)$. Hence $a \leq 2s-3n-3+k_H$. In view of $2s \leq 3n+2$, it follows that $a \leq k_H-1$. Since H meets k_H other lines which are parallel to L, the assertion follows.■

<u>6.3 Lemma</u>. Suppose L and G are disjoint lines of degree $n-1$. If $\mathcal{M}$ is the set of long lines $\neq$ G which miss L and meet G, then $\mathcal{M}$ is a clique with $|\mathcal{M}| \geq n-1-\alpha$.

<u>Proof</u>. Suppose that L_1, L_2, L_3 are mutually intersecting lines which miss L and denote the degree of L_j by $n+1-d_j$. Then $d_1d_2+d_1d_3+d_2d_3-d_1-d_2-d_3 \geq n$ by Lemma 3.7 a). In the case that $d_1 = 2$, we obtain $(d_2+1)(d_3+1) \geq n+3$.

In view of $(e+1)^2 \leq n+2$ and because every long line has degree at least $n+1-e$, it follows that $\mathcal{M}$ is a clique. Since every point of G lies on a second line parallel to L and in view of Lemma 6.2 a), we have $|\mathcal{M}| \geq k_G-\alpha = n-1-\alpha$.■

<u>6.4 Lemma</u>. Let $\mathcal{M}_1$ and $\mathcal{M}_2$ be distinct cliques.

a) If $H \in \mathcal{M}_1$ and $K \in \mathcal{M}_2$ with $|H \cap K| = 1$, then $|\mathcal{M}_1 \cap \mathcal{M}_2| \leq d_H d_K$.

b) If $H,K \in \mathcal{M}_1 \cap \mathcal{M}_2$ with $H \neq K$, then $|\mathcal{M}_1 \cup \mathcal{M}_2| \leq n+1+(d_H-1)(d_K-1)$.

<u>Proof</u>. a) Every line contained in both cliques is parallel to H and to K. Since the number of lines parallel to H and K is $d_H d_K$, the assertion follows.

b) Every line $\neq$ H,K of one of both cliques is parallel to H and K. Because H and K are parallel, the number of lines parallel to H and K is $n-1+(d_H-1)(d_K-1)$. This implies b).■

<u>6.5 Lemma</u>. Suppose that L is a line of degree $n-1$. Then there are exactly two good cliques containing L. Every long line missing L is contained in exactly one of these two cliques.

<u>Proof</u>. Denote by $2n-\beta$ the number of long lines which miss L. By Lemma 6.2, $\beta \leq \alpha$. Lemma 6.2 b) shows that there are two intersecting lines L_1 and L_2 of degree $n-1$ which miss L. Let $\mathcal{M}_1'$ be the set consisting of L and the long lines $\neq L_2$ which meet L_2 and miss L, and let $\mathcal{M}_2'$ be the set consisting of L and the long lines $\neq L_1$ which meet L_1 and miss L. Lemma 6.3 shows that $\mathcal{M}_1'$ and $\mathcal{M}_2'$ are cliques. Since L_j meets $n-1$ other lines which miss L, we have $|\mathcal{M}_1'| \geq n-\beta$ and $|\mathcal{M}_2'| \geq n-\beta$. Let $\mathcal{M}_1$ and $\mathcal{M}_2$ be maximal cliques with $\mathcal{M}_1' \subseteq \mathcal{M}_1$ and $\mathcal{M}_2' \subseteq \mathcal{M}_2$. We claim that $\mathcal{M}_1$ and $\mathcal{M}_2$ have only the line L in common.

Assume to the contrary that $\mathcal{M}_1$ and $\mathcal{M}_2$ have a line $H \neq L$ in common. Then Lemma 6.4 b) shows that $|\mathcal{M}_1 \cup \mathcal{M}_2| \leq n+1+(d_L-1)(d_H-1) \leq n+e$. Lemma 6.4 a) used for L_1 and L_2 shows that $|\mathcal{M}_1 \cap \mathcal{M}_2| \leq 4$. Hence,

$$2(n-\alpha) \leq 2(n-\beta) \leq |\mathcal{M}_1|+|\mathcal{M}_2| = |\mathcal{M}_1 \cup \mathcal{M}_2|+|\mathcal{M}_1 \cap \mathcal{M}_2| \leq n+4+e \leq n+3e,$$

which contradicts (5). Consequently $\mathcal{M}_1 \cap \mathcal{M}_2 = \{L\}$.

We want to show that $\mathcal{M}_1$ and $\mathcal{M}_2$ are the only maximal cliques $\mathcal{M}$ satisfying $|\mathcal{M}| \geq n-\alpha$ and $L \in \mathcal{M}$. Assume to the contrary that there exists a third maximal clique $\mathcal{M}$ with $|\mathcal{M}| \geq n-\alpha$ and $L \in \mathcal{M}$. Since $\mathcal{M}_1$ and $\mathcal{M}_2$ have even $n-\beta$ lines, we have

$$\begin{aligned} |\mathcal{M}-\{L\}|+|(\mathcal{M}_1 \cup \mathcal{M}_2)-\{L\}| &= |\mathcal{M}|+|\mathcal{M}_1|+|\mathcal{M}_2|-3 \\ &\geq 3n-\alpha-2\beta-3 \geq 3n-3-2\alpha-\beta > 2n-\beta. \end{aligned}$$

Since $2n-\beta$ is the number of long lines missing L, it follows that there exists a line $G \neq L$ which is contained in $(\mathcal{M}_1 \cup \mathcal{M}_2) \cap \mathcal{M}$. We may assume that $G \in \mathcal{M}_2 \cap \mathcal{M}$. Lemma 6.4 b) shows then that $|\mathcal{M}_2 \cup \mathcal{M}| \leq n+1+(d_L-1)(d_G-1) = n+d_G \leq n+e$.

If $L_2 \notin \mathcal{M}$, then L_2 meets a line X of $\mathcal{M}$, since $\mathcal{M}$ is a maximal clique. In this case, Lemma 6.4 a) shows that $|\mathcal{M}_2 \cap \mathcal{M}| \leq 2d_X$. Since $|\mathcal{M}|+|\mathcal{M}_2| = |\mathcal{M} \cup \mathcal{M}_2|+|\mathcal{M} \cap \mathcal{M}_2|$, it follows that $2(n-\alpha) \leq |\mathcal{M}|+|\mathcal{M}_2| \leq n+e+2d_X \leq n+3e$, contradicting (5).

Suppose now that $L_2 \in \mathcal{M}$. Since the number of lines missing L_1 and L_2 is four, it follows that L_1 meets at least $|\mathcal{M}|-4$ of the lines of $\mathcal{M}$. By construction, these lines belong to $\mathcal{M}_2$. Furthermore, $L \in \mathcal{M}_2 \cap \mathcal{M}$. Hence, $|\mathcal{M}_2 \cap \mathcal{M}| \geq |\mathcal{M}|-3 \geq n-3-\alpha$. Since $\mathcal{M}$ and $\mathcal{M}_2$ are distinct maximal cliques, there are lines $H \in \mathcal{M}$ and $H_2 \in \mathcal{M}_2$ which meet each other. Lemma 6.4 a) shows therefore that $|M_2 \cap M| \le e^2$. It follows that $n-3-\alpha \le e^2$, which contradicts (5).

Consequently, M_1 and M_2 are the only maximal cliques which contain L and have at least $n-\alpha$ lines. If H is any long line missing L, then Lemma 6.2 shows that there is a line G of degree $n-1$ which meets H and misses G. Lemma 6.3 shows therefore that there is a maximal clique which contains H and L and which has at least $n-\alpha$ lines. Hence $H \in M_1$ or $H \in M_2$ for every long line H missing L. Since L misses at least $2n-\alpha$ long lines and in view of $L \in M_j$, it follows that $|M_1|+|M_2| \ge 2n+2-\alpha$. On the other hand, we have $|M_j| \le n+1$, since a line of degree $n+1$ meets every other line (each point lies on a line of degree $n+1$, since there are no n-lines!). Hence $|M_j| \ge n+1-\alpha$, which proves that M_1 and M_2 are good cliques.■

<u>Remark</u>. Suppose we replace $|M| \ge n+1-\alpha$ by $|M| \ge n-\alpha$ in the definition of "good clique" and suppose that

(5') $2(n-\alpha) > n+1+(e-1)^2+e^2$.

Then Lemma 6.4 shows that every two distinct good cliques have at most one line in common. Suppose furthermore that $3(n-1-\alpha) > 2n$. Then it follows that every line of degree $n-1$ is contained in at most two good cliques. With this, Lemma 6.5 could be proved much easier and also Lemma 6.6 would already be proved. However, we would get a little weaker bound for the embedding of $(n+1,1)$-designs in Corollary 6.12. This is the reason why we accept some difficulties.

<u>6.6 Lemma</u>. If M_1 and M_2 are distinct good cliques, then M_1 and M_2 have at most one line in common.

<u>Proof</u>. Let L_j be a line of degree $n-1$ of M_j. Since M_1 and M_2 are different maximal cliques, there are lines $K \in M_1$ and $H \in M_2$ which intersect each other.

Assume by way of contradiction that M_1 and M_2 have two lines X and Y in common. Then Lemma 6.5 shows that $M_1 \cap M_2$ contains no line of degree $n-1$. Hence, $L_1 \notin M_2$ and $L_2 \notin M_1$. Lemma 6.4 implies that

$$2(n+1-\alpha) \le |M_1|+|M_2| = |M_1 \cup M_2|+|M_1 \cap M_2| \le n+1+(d_X-1)(d_Y-1)+d_K d_H.$$

If L_1 and L_2 intersect, then we can choose $K = L_1$ and $H = L_2$ and obtain $2(n+1-\alpha) \leq n+5+(d_X-1)(d_Y-1) \leq n+5+(e-1)^2$. This contradicts (5).

If L_1 and L_2 are parallel, let M be the unique good clique containing L_1 and L_2. Then $M \neq M_1, M_2$. By Lemma 6.2, L_2 intersects at least $n-1-\alpha$ long lines which are parallel to L_1. These long lines are not contained in M (since $L_2 \in M$). Consequently, they are contained in M_1 (Lemma 6.5). This shows that there are at least $n-\alpha$ lines in M_1 which are not lines of M_2 (also L_1 is in M_1-M_2). In the same way follows that $|M_2-M_1| \geq n-\alpha$. Together with Lemma 6.4 b), we obtain

$$2(n-\alpha) \leq |M_1 \cup M_2| \leq n+1+(d_X-1)(d_Y-1) \leq n+1+(e-1)^2,$$

and hence $n \leq e^2-2e+2+2\alpha$, which contradicts (5).■

<u>6.7 Lemma</u>. a) There exists a point which lies only on long lines. For any such point p, every good clique contains a line through p.

b) If L and H are parallel long lines, then there are at most d_L-1 good cliques containing L but not H.

c) If L is a long line, then there are at most d_L good cliques containing L.

d) The number of good cliques is at most s with equality if and only if every long line L is contained in exactly d_L good cliques.

<u>Proof</u>. a) Let q be any point. In view of $v \geq n^2-n$, q lies on a line L of degree at least n. By assumption (1) of Proposition 6.1, L contains a point lying only on long lines. Denote by p any point lying only on long lines.

Let M be any good clique and L a line of degree $n-1$ of it. We have to show that M contains a line passing through p. For this, we may assume that $p \notin L$. If L_1 and L_2 are the two lines through p which are parallel to L, then Lemma 6.5 shows that there is a unique good clique M_j, $j = 1,2$, which contains L and L_j. Since $M_1 \neq M_2$, Lemma 6.5 implies also that M is one of these two cliques. This shows that M contains L_1 or L_2, which proves a).

b) By (1) of Proposition 6.1, there is a point p on H which lies only on long lines. By a), every clique contains a line passing through p, and, by Lemma 6.6, for every line X through p which is parallel to L there is at most one good clique containing L and X. Since p is contained in exactly d_L lines parallel to L and because H is one of these lines, this proves b).

c) Obviously, there is no good clique containing L, if L has degree n+1. Suppose now that $k_L \leq n-1$. As in Lemma 6.2, we can prove that there is a long line H missing L. Lemma 6.6 shows that there is at most one clique containing H and L. Part c) follows therefore from Part b).

d) Let p be a point which is contained only in long lines. If $L_1,\ldots,L_{n+1}$ are the lines through p, and if the degree of L_j is $n+1-d_j$, then $s = d_1+\ldots+d_{n+1}$. Therefore, a) and c) show that there are at most s good cliques, and that there are exactly s good cliques, if each of the long lines L_j is contained in d_j cliques.

Now suppose the number of good cliques is s, and let L be any long line. In view of the assumption (1) in Proposition 6.1, we may assume that p is a point of L so that the preceding paragraph shows that L is contained in d_L good cliques. This proves d).■

6.8 Lemma. If $e \geq 3$, then every line of degree n-2 is contained in three good cliques.

Proof. Suppose that L is a line of degree n-2. Lemma 3.1 c) yields that the number of lines of degree n-1 which are parallel to L is at least $6n+9-3s \geq n+3$. Therefore there exists two intersecting lines L_1 and L_2 of degree n-1 which are parallel to L. Let M_j be the good clique containing L and L_j (Lemma 6.5). If there is a line of degree n-1 which is parallel to L and which is not contained in $M_1 \cup M_2$, then it follows from Lemma 6.5 that L is contained in a third good clique. We may therefore assume that every line of degree n-1 which is parallel to L is contained in $M_1 \cup M_2$. Since every clique has at most n+1 lines, this implies that at least n of the lines parallel to L have degree at most n-2.

Denote by w the number of lines of degree at most n+1-e which are parallel to L. Because at most 2n of the lines parallel to L have degree n-1, Lemma 3.1 c) shows that

$$3(n^2+3-s) \leq 2n(n-1) + w(n+1-e) + (n-w)(n-2).$$

It follows that $w(e-3) \leq 3s-4n-9$.

Let X be the set of lines parallel to L which intersect L_1 and L_2, and denote by Y the subset of X consisting of the lines of X which have degree at least n+2-e. By Lemma 3.7 b), $|X| \geq n-4$. Hence $y := |Y| \geq n-4-w$. It follows from Lemma 3.7 a) that Y is a clique.

Assume by way of contradiction that $2y \leq s-3$. Then

$$2(n-4)(e-3)-2(3s-4n-9) \leq 2(n-4-w)(e-3) \leq 2y(e-3) \leq (s-3)(e-3),$$

and hence $n(2e+2)-5e+33 \leq s(e+3)$, which contradicts hypothesis (6) of Proposition 6.1. Consequently $2y \geq s-2$.

Assume by way of contradiction that M_1 and M_2 are the only good cliques containing L. Then Lemma 6.7 d) shows that there are at most s-3 good cliques different from M_1 and M_2.

As we have shown that L is contained in two good cliques, we can show that every long line is contained in at least two good cliques. In particular, each of the y lines of Y is contained in two good cliques, which are different from M_1 and M_2. In view of $2y > s-3$, this implies that there is a good clique M which contains two lines K and H of Y. Lemma 6.4 b) used for the cliques M and $Y \cup \{L\}$ shows that $|M \cup Y \cup \{L\}| \leq n+1+(d_K-1)(d_H-1) \leq n+1+(e-2)^2$.

By our hypothesis, $L \notin M$. Let G be any line of degree n-1 of M. If G intersects L, then L intersects at least $|M|-6 \geq n-5-\alpha$ of the lines of M, because there are only 6 lines parallel to L and G. In this case, we have therefore $m := |M-Y| \geq n-5-\alpha$. If G is parallel to L, let N be the good clique containing G and L. Since $L \notin M$, we have $N \neq M$. Lemma 6.2 a) shows that there are at least $k_L-\alpha = n-2-\alpha$ long lines which are parallel to G and intersect L, and Lemma 6.5, used for G, yields that these lines belong to M (since they are not lines of N). This proves that $m \geq n-2-\alpha \geq n-5-\alpha$ in this case. It follows that

$$2n-8-\alpha-w \leq n-4-\alpha+y \leq m+y+1 \leq |M \cup Y \cup \{L\}| \leq n+1+(e-2)^2.$$

Hence $n \leq e^2-4e+13+\alpha+w$. Using $n \geq e^2-2e+5+2\alpha$ (hypothesis (5) of Proposition 6.1), we obtain $\alpha+1 \leq w-2e+9$. Now the definition of α shows that

$$2(s-2-n) + 1 \leq (\alpha+1)(e-1) \leq w(e-1)-(2e-9)(e-1).$$

In view of $w(e-3) \leq 3s-4n-9$, we obtain

$$(2s-3-2n)(e-3) \leq [w(e-1)-(2e-9)(e-1)](e-3)$$
$$\leq (3s-4n-9)(e-1)-(2e-9)(e-1)(e-3),$$

i.e. $2(e+1)n+6e+(2e-9)(e-1)(e-3) \leq s(e+3)$, which contradicts $e \geq 3$ and (6). This final contradiction proves that L is contained in three good cliques.■

6.9 Lemma. a) If L is a long line, then L is contained in precisely d_L good cliques, and every long line which is parallel to L is contained in exactly one of these d_L cliques.

b) The number of good cliques is s.

Proof. a) We proceed by induction on $d := d_L$. If $d = 2$, then the assertion follows from Lemma 6.5.

Let us now suppose that $d \geq 3$. We want to show first that L lies in at least d cliques. If $d = 3$, then this follows from Lemma 6.8. If $d \geq 4$, let x denote the number of lines parallel to L which have degree at least $n+2-d$. Then Lemma 3.1 b) implies that

$$d(n^2+d-s) \leq x(n-1) + (dn-x)(n+1-d) = dn(n+1-d) + x(d-2).$$

so that $x > (d-1)n$. By the induction hypothesis, each of these x lines is contained in a unique good clique which contains also L. Since every clique has at most $n+1$ lines, it follows therefore that L is contained in d cliques.

Consequently, a line of degree $n+1-d$, $2 \leq d \leq e$, is contained in at least d good cliques. Now, Lemma 6.7 b) and c) and Lemma 6.6 show that L is contained in precisely d good cliques and that every long line parallel to L is contained in exactly one of these good cliques.

b) follows from part a) and Lemma 6.7 d).■

6.10 Lemma. L can be embedded into a projective plane of order n.

Proof. Let L' be the partial linear space consisting of the points and long lines of L. From L' we obtain a new partial linear space D, if we let the lines of each good clique M intersect in a new point p_M. More precisely, if p_M is a new symbol for each good clique M, then $D = (p \cup p', \{L^* \mid L \text{ is a long line}\})$ where

$$L^* = L \cup \{p_M \mid M \text{ is a good clique containing } L\},$$

for each long line L, and

$$p' = \{p_M \mid M \text{ is a good clique of } L\}$$

In view of Lemma 6.9 b), D has n^2+n+1 points. Lemma 6.9 a) shows that every line of D has degree n+1, and that any two lines of D intersect. Furthermore, since L has at most n short lines, D has at least n^2+1 lines. Hence, if D_d is the structure dual to D, then D_d is an (n+1,1)-design with n^2+n+1 lines and at least n^2+1 points. Corollary 3.5 a) shows that D_d can be embedded into a projective plane of order n. Consequently, also D can be embedded into a projective plane P of order n. Since L' lies in D, L' can be embedded into P. Now, Lemma 3.8 shows that L can be embedded into the projective plane P.■

Lemma 6.10 completes the proof of Proposition 6.1. Now we state the main result of this section.

6.11 Theorem. Let L be an (n+1,1)-design with n^2+n+1 lines for some integer $n \geq 7$, and denote the number of points by $n^2+n+1-s$. Suppose furthermore that there are integers e_1 and e_2 with $2 \leq e_1 \leq e_2$ and $(e_2+1)^2 \leq n+2$ such that every integer e with $e_1 \leq e \leq e_2$ satisfies the following property (P):

(P) Let α be the maximal integer with $\alpha(e-1) \leq 2(s-2-n)$. Then

(i) If $e = 2$, then $n \geq 8+\alpha$ and $n \geq 7+2\alpha$.

(ii) If $e = 3$, then $n \geq 13+\alpha$, $n \geq 10+2\alpha$ and $8n+17 \geq 6s$.

(iii) If $e = 4$, then $10n+12 \geq 7s$.

(iv) If $e = 5$, then $2s \leq 3n+1$.

(v) If $e \in \{4,5\}$ and $n \leq 36$, then $n \geq e^2-2e+5+2\alpha$.

(vi) If $4 \leq e \leq 7$ and $n \leq 62$, then $n \geq e^2+4+\alpha$.

If furthermore $2s \leq 3n+2$ and

$$(+) \qquad (n-e_2)(e_2^2-1) + \sum_{e=e_1}^{e_2} (e^2-1) > s(s-2-n),$$

then L can be embedded into a projective plane of order n.

Proof. We proceed by induction on s. If $s = 0$, then L itself is a projective plane. If $s > 0$ and L has a line of degree n, then L possesses a parallel class so that we may adjoin an infinite point to complete the induction.

Let us now suppose that $s > 0$ and that there is no line of degree n. Then Lemma 3.3 b) and (+) imply that for some integer e with $e_1 \leq e \leq e_2$ there are at most $n-e$ lines of degree at most $n-e$. Hence, e satisfies the conditions (1) to (4) of Proposition 6.1. Now Proposition 6.1 completes our proof, if we show that e satisfies also (5) and (6) of Proposition 6.1.

Let α be the maximal integer with $\alpha(e-1) \leq 2(s-2-n)$. Then (5) and (6) follows directly from (P)(i) and (ii), if $e \leq 3$. Let us suppose $e \geq 4$ from now on. Notice that $\alpha(e-1) \leq 2(s-2-n) \leq n-2$.
Assume by way of contradiction that $n \leq e^2-2e+4+2\alpha$. Then

$$n(e-1) \leq (e^2-2e+4)(e-1)+2\alpha(e-1) \leq (e^2-2e+4)(e-1)+2n-4,$$

i.e. $n(e-3) \leq (e^2-2e+4)(e-1)-4$. In view of $n \geq (e_2+1)^2-2 \geq (e+1)^2-2 \geq e^2+2e-1$, we conclude that $e \in \{4,5\}$ and that $n \leq 36$. This contradicts (v). Consequently, $n \geq e^2-2e+5+2\alpha$. Since $e \geq 4$, this implies that $n \geq 3e+1+2\alpha$.
Assume by way of contradiction that $n \leq e^2+3+\alpha$. Then

$$n(e-1) \leq (e^2+3)(e-1)+\alpha(e-1) \leq (e^2+3)(e-1)+n-2.$$

In view of $n \geq e^2+2e-1$, we conclude that $e \leq 7$ and $n \leq 62$. This contradicts (vi). Consequently $n \geq e^2+4+\alpha$, so that condition (5) of Proposition 6.1 holds.
Condition (6) of 6.1 follows from (iii) and (iv), if $e \leq 5$, and from $2s \leq 3n+2$ and $n \geq e^2+2e-1$, if $e \geq 6$. This completes the proof of Theorem 6.11.■

Now we are ready to give bounds for the embedding of (n+1,1)-designs with n^2+n+1 lines.

6.12 Corollary. Let L be an (n+1,1)-design with n^2+n+1 lines for some integer n. Then L can be embedded into a projective plane of order n, if one of the following conditions is satisfied:

(a) $v \geq n^2-\frac{1}{2}n$, $n \geq 223$, and $n \neq 252$.

(b) $v \geq n^2-\frac{1}{2}n+1$ and $n = 252$.

<u>Proof</u>. Let e be the unique integer with $(e+1)^2-2 \le n < (e+2)^2-2$. Suppose that $n \ge 223$ so that $e \ge 14$. Suppose furthermore that $v \ge n^2-\frac{1}{2}n$ and set $s = n^2+n+1-v$. In view of $n > 62$, Theorem 6.11 shows that L can be embedded, if

$$\text{(A)} \quad (n-e)(e^2-1) + \sum_{e=6}^{e_2} (d^2-1) > s(s-2-n).$$

In view of $2s \le 3n+2$, condition (A) is fulfilled, if

$$\text{(B)} \quad 2e(e+1)(2+1)/3-220 = \sum_{d=6}^{e} 4(d^2-1) > (3n+2)(n-2) - 4(n-e)(e^2-1).$$

For a fixed value for $e \ge 14$, the right hand side of (B) increases with n. It is easily checked that (B) is satisfied for $e \ge 15$ and $n = (e+2)^2-3$, which is the maximal possibile value for n (e beeing fixed). If $e = 14$, then the maximal possible value for n is $(e+2)^2-3 = 253$. However, (B) is satisfied only for $n = 251$ and hence for every value of n with $223 = (e+1)^2-2 \le n \le 251$. If $n = 253$, then $2s \le 3n+2$ implies that $s \le (3n+1)/2 = 378$ and condition (A) is fulfilled. If $n = 252$, then condition (A) is only fulfilled for $s \le 3n/2 = 378$, i.e. for $v \ge n^2+\frac{1}{2}n+1$.■

<u>6.13 Remark</u>. We have seen that (n+1,1)-desings with $v \ge n^2-\frac{1}{2}n$ points and n^2+n+1 lines can be embedded for $n \ge 253$. Running a computer program for small values of n, we can find the minimal non-negative integer c such that Theorem 6.11 proves the embeddability of (n+1,1)-designs with n^2+n+1 lines and $v \ge n^2-\frac{1}{2}n+c$ points. In the following list we give all obtained pairs (n,c) with $7 \le n \le 252$ for which $c > 0$.

n	c	n	c	n	c	n	c	n	c	n	c
7	3	8	3	9	3	10	4	11	3	12	4
13	4	14	5	15	4	16	4	17	4	18	4
19	4	20	4	21	4	22	5	23	5	24	4
25	3	26	3	27	3	28	4	29	4	30	5
31	5	32	5	33	5	34	5	35	3	36	3
37	3	38	3	39	3	40	4	41	4	42	5
43	5	44	6	45	6	46	6	47	3	48	2
49	2	50	3	51	3	52	3	53	3	54	4
55	4	56	5	57	5	58	6	59	6	60	6
61	6	62	1	63	1	64	2	65	2	66	3
67	2	68	3	69	3	70	4	71	4	72	5

n	c	n	c	n	c	n	c	n	c	n	c
73	4	74	5	75	5	76	6	77	6	78	7
80	1	82	1	83	1	84	2	85	2	86	3
87	3	88	3	89	3	90	4	91	4	92	5
93	5	94	6	95	6	96	6	97	6	102	1
104	1	105	1	106	2	107	2	108	3	109	3
110	3	111	3	112	4	113	4	114	5	115	5
116	6	117	6	118	6	128	1	129	1	130	2
131	2	132	2	133	2	134	3	135	3	136	4
137	4	138	5	139	5	140	5	141	5	154	1
156	1	157	1	158	2	159	2	160	3	161	3
162	3	163	3	164	4	165	4	166	5	184	1
186	1	187	1	188	2	189	2	190	3	191	3
192	3	193	3	218	1	219	1	220	2	221	2
222	2	252	1								

Remark 6.13 and Corollary 6.12 now give the following result in which there is no longer a restriction on n (see Corollary 3.5 for $n \le 6$).

<u>6.14 Corollary</u>. Let L be a linear space with constant point degree n+1 and n^2+n+1 lines for some integer $n \ge 2$. If L has $v > n^2-\frac{1}{2}n+6$ points, then L can be embedded into a projective plane of order n.■

Up to now, we tried to find an integer c such that $v > n^2-\frac{1}{2}n+c$ implies embeddability for all integers $n \ge 2$. We accepted some difficulties to find a small value for c and obtained $c = 6$. The reason why we wanted this small value for c is that we shall also classify all linear spaces with $n^2-\frac{1}{2}n+1 < v \le b$ and not constant point degree n+1 in the next section. For large values of n, the proof would have been much easier, we can even improve the bound for v then:

<u>6.15 Theorem</u>. Let L be an (n+1,1)-design with n^2+n+1 lines, and denote the number of points of L by $n^2+n+1-s$. If $n \ge 2$ and $2s \le (1+\sqrt{5})n - 34\sqrt{n}/\sqrt{5}$, then L can be embedded into a projective plane of order n.

Again, this result could be improved a little (but only the coefficient of $\sqrt{n}$), if you accept some difficulties. However, in order to make the proof clearer and shorter, we avoid difficulties this time.

As before, the theorem will be proved in several lemmata. If L has a line of degree n, then we may (as in the proof of Theorem 6.11) adjoin an infinite point to L and apply induction. We may therefore assume that there is no line of degree n. Furthermore, since Theorem 6.15 follows from Corollary 6.14, if $n < 3900$, we shall assume that $n \geq 3900$.

Let e be the unique positive integer with $(e+7)^2 < n \leq (e+8)^2$ and denote by α the maximal integer with $\alpha(e-1) \leq 5(s-2-n)$. Call a line *short*, if it has degree at most $n-e$, and otherwise *long*. A *clique* is a set of mutually parallel long lines, and a clique M is called *good*, if $|M| \geq n-27-\alpha$ and if M contains a line of degree at least $n-4$.

<u>6.16 Lemma</u>. a) $n > e^2+4e+53+2\alpha$, $(n+1-e)(e^2-1) > s(s-2-n)$ and $n > 169+6\alpha$.
b) There are at most $n-e$ short lines.

<u>Proof</u>. a) In view $n \geq 3900$, we have $e \geq 54$. Therefore it is easy to see that a) follows from $(e+7)^2 < n \leq (e+8)^2$, the bound for s in 6.15, and the definition of α.

b) follows from Lemma 3.3 b) and $(n+1-e)(e+1)(e-1) > s(s-2-n)$.■

<u>6.17 Lemma</u>. If L is a line of degree at least $n-4$, then there are at most α short lines parallel to L.

<u>Proof</u>. This follows from Lemma 3.1 in the same way as we proved Lemma 6.2.■

<u>6.18 Lemma</u>. Any two different good cliques $\mathcal{M}_1$ and $\mathcal{M}_2$ have at most one line in common.

<u>Proof</u>. Since $\mathcal{M}_1$ and $\mathcal{M}_2$ are distinct maximal cliques, there exists a line $H \in \mathcal{M}_1-\mathcal{M}_2$. Again, since $\mathcal{M}_2$ is maximal, H intersects some line G of $\mathcal{M}_2$.

Assume by way of contradiction that $\mathcal{M}_1$ and $\mathcal{M}_2$ have two lines K and L in common. As in Lemma 6.4, this implies that

$$\begin{aligned} 2n-54-2\alpha \leq |\mathcal{M}_1|+|\mathcal{M}_2| &= |\mathcal{M}_1\cup\mathcal{M}_2|+|\mathcal{M}_1\cap\mathcal{M}_2| \\ &\leq n+1+(d_K-1)(d_L-1)+d_H d_G. \end{aligned}$$

Let X be any line of degree at least $n-4$ of $\mathcal{M}_1$. If $X \notin \mathcal{M}_2$, then we can choose $H = X$, and if $H \in \mathcal{M}_2$, then we can choose $K = X$. In both cases we obtain $n \leq e^2+4e+53+2\alpha$, a contradiction.■

<u>6.19 Lemma</u>. Suppose that L is a line with $2 \le d := d_L \le 5$. Then there are exactly d good cliques containing L, and every long line parallel to L is contained in one of them.

<u>Proof</u>. Let H be any long line parallel to L. For every point p of H, we define an integer a_p in the following way. If $H_2,\ldots,H_d$ are the lines different from H which pass through p and which miss L, then set $a_p := |H_2|+\ldots+|H_d|$. Because the $dn-k_H(d-1)$ lines parallel to L which do not meet H in one point have degree at most $n-1$, Lemma 3.1 shows that

$$d(n^2+d-s) \le [nd-k_H(d-1)](n-1) + \sum_{p\in H} a_p .$$

In view of $3s \le 5n-3\sqrt{n}$, we obtain

$$\sum_{p\in H} a_p \ge k_H(d-1)(n-1)-d(s-n) \ge k_H(d-1)(n-1)-d(\tfrac{2}{3}n-\sqrt{n}).$$

Consider first the case $d \ge 3$. Then

$$d(\tfrac{2}{3}n-\sqrt{n}) < d(\tfrac{2}{3}n+1-e) < (d-1)(n+1-e) < k_H(d-1).$$

Thus, there is a point $q \in H$ with $a_q \ge (d-1)(n-1)+2-d$. Let $L_2,\ldots,L_d$ be the lines other than H through q which are parallel to L and denote the degree of L_j by $n+1-d_j$. Set $D := d_2+\ldots+d_d$ and

$$A := \sum_{2\le j<k} d_j d_k.$$

Then $D = (d-1)(n+1)-a_q \le 3d-4$. Since $2 \le d \le 5$ and $d_j \ge 2$, it follows that $A \le 45$.

Let N denote the set of lines parallel to L which meet each of the lines L_j in one point, and let M be the set consisting of the long lines of N. By Lemma 3.7 b), $|N| \ge n-(d-1)(D+1-d) \ge n-4\cdot 7 \ge n-28$, so Lemma 6.17 shows $|M| \ge n-28-\alpha$.

Assume by way of contradiction that there exists lines L_0 and L_1 in M which meet each other, and set $|L_j| = n+1-d_j$. Then Lemma 3.7 a) shows that

$$n \le d_0d_1 + (d_0+d_1)D + A - (d-1)(d_0+d_1+D) + \tfrac{1}{2}(d-2)(d+1).$$

Together with $d_0,d_1 \le e$, $d \le 5$, $A \le 45$ and $D \le 3d-4$, we obtain

$$\begin{aligned} n &\le e^2 + 2e(D+1-d) + 45 - (d-1)D + 9 \\ &\le e^2 + 2e(2d-3) - (d-1)(2d-3) + 54 \le (e+7)^2 - 39, \end{aligned}$$

which contradicts the definition of e.

Consequently, $\mathcal{M}$ is a clique of long lines. Since $|\mathcal{M}| \geq n-28-\alpha$, it follows that $\{L\} \cup \mathcal{M}$ is contained in a good clique $\mathcal{M}_1$. By the construction of $\mathcal{M}_1$, the lines H and L are in $\mathcal{M}_1$.

Now consider the case d = 2. Then

$$d(\tfrac{2}{3}n-\sqrt{n}) < 2k_8.$$

Hence, H contains a point q with $a_q \geq n-2$, i.e. H intersects a line H_2 of degree at least n−2 which is parallel to L. If $\mathcal{M}$ denotes the set of long lines which miss L and meet H_2, then it follows as before from Lemma 3.7 a) that $\mathcal{M}$ is a clique. Since H_2 has degree at least n−2, we have $|\mathcal{M}| \geq n-2-\alpha$. Consequently, $\{L\} \cup \mathcal{M}$ is contained in a good clique $\mathcal{M}_1$, which contains H and L.

We have shown that for every long line X parallel to L there is good clique containing X and L. Let $\mathcal{M}_1,\dots,\mathcal{M}_t$ be the good cliques containing L. By Lemma 6.18 any two of these cliques have just the line L in common. It remains to show that t = d. Since L is parallel to dn lines, since every good clique has at least $n-27-\alpha$ lines, and in view of $(d+1)(n-28-\alpha) > dn$, we have $t \leq d$. Let p be a point outside L which is contained only in long lines (such a point exists in view of Lemma 6.16 b)). Then each of the d long lines through p which is parallel to L lies in one of the cliques $\mathcal{M}_j$. This shows that $t \geq d$.■

<u>6.20 Lemma</u>. If L is a long line, then there are exactly d_L good cliques containing L and every long line parallel to L is contained in exactly one of them. Furthermore, the number of good cliques is s.

<u>Proof</u>. We can prove the assertions of Lemma 6.7 by applying almost the same proof. The only difference is that a good clique now need not to contain a line of degree n−1 but only a line of degree at least n−4. Therefore it suffices to show that every line L of degree n+1−d, $2 \leq d \leq e$, lies in at least d good cliques.

We proceed by induction on d. If $d \leq 5$, then Lemma 6.20 follows from Lemma 6.19. Now suppose that $d \geq 6$ and denote by a the number of lines parallel to L which have degree at most n+1−d. Lemma 3.1 yields

$$d(n^2+d-s) \leq (dn-a)(n-1)+a(n+1-d) \leq dn(n-1) - a(d-2).$$

It follows that $3a(d-2) \leq 3d(s-n-d) < 2dn$, and hence $a < n$. Thus, at least $(d-1)n+1$ of the lines parallel to L have degree at least $n+2-d$. The induction hypothesis shows that for each of these lines X, there is a good cliques containing X and L. Because every good clique can have at most $n+1$ lines, we conclude that L is contained in at least d good cliques.■

As in the proof of Lemma 6.10, it follows from Lemma 6.20 that L can be embedded into a projective plane of order n, and this completes the proof of Theorem 6.15.

Theorem 6.15 improves Corollary 6.14 only for large n and Corollary 6.14 only considers the case $n \geq 7$. What about $n \leq 6$? The best result we know up to now is that $v \geq n^2$ implies embeddability. With our methods, we can even not improve this result. The case $v = n^2-1$ is particularly difficult if every line has degree $n-1$ or $n+1$. This is the pseudo-complement of a hyperoval ($(n+2)$-arc) in a projective plane of even order. We know that it can be embedded, if $n > 7$ (see Remark 6.13). Also the cases $n = 7$ and $n < 6$ are easy:

<u>6.21 Lemma</u> (Bose and Shrikhande, 1973). Suppose L is an $(n+1,1)$-design with n^2+n+1 lines and $n^2-1 \geq 3$ points. Then one of the following cases occurs:
(1) L can be embedded into a projective plane of order n.
(2) $n = 6$, and every line has degree $n-1$ or $n+1$.

<u>Proof</u>. In view of Corollary 3.5, we may assume that no line has degree n. Furthermore, since $v \neq n^2+n+1$, not every line has degree $n+1$. Let L be any line of degree $n+1-d \leq n-1$. Since L is parallel to dn lines, and because every line parallel to L has degree at most $n-1$, Lemma 3.1 c) implies that $d = 2$. Hence every line has degree $n-1$ or $n+1$. In view of $v = n^2-1$, it follows that every point is contained in $1+\frac{1}{2}n$ lines of degree $n-1$ and in $\frac{1}{2}n$ lines of degree $n+1$. In particular, n is even. If $n = 6$, then (2) occurs. We have therefore to show that L can be embedded, if $n \neq 6$. For $n = 2$ this is trivial and for $n \geq 8$ it follows from Remark 6.13.

Consider finally the case $n = 4$. Then every point is contained in two lines of degree $n+1 = 5$. Since any two 5-lines meet, the number of 5-lines is six and the structure D consisting of all $n^2-1=15$ points and the six 5-lines is uniquely determined. It is dual to the complete graph C_6 on six points. If you take two parallel lines of C_6, then there is a unique third line which is parallel to the given two lines. Hence if you take two unjoined points p and q of D, then there is a unique third point which is unjoined to p and q. This shows that a 3-line of L is already determined by D and two of its points. Hence, L is uniquely determined by D so that L is uniquely determined up to isomorphism. Thus, if L' is the complement of a hyperoval in the projective plane of order 4, then L and L' are isomorphic. This completes the proof.■

Remarks. 1) De Witte (1977a) showed that there is no (7,1)-design which meets the condition (2) of Lemma 6.21. The proof uses a characterization theorem of Li-chien (1960) of certain partially balanced incomplete block designs. We shall give a self-contained proof of de Witte's theorem in section 13.

2) Lemma 6.21 is a special case of a theorem of Thas and De Clerck (1975).

6.22 Lemma (McCarthy and Vanstone, 1977). Every (n+1,1)-design with n^2-2 points and n^2+n+1 lines, $n \geq 2$, can be extended to an (n+1,1)-design with n^2-1 points and n^2+n+1 lines.

Proof. In view of Lemma 3.4 and Lemma 6.22, it suffices to show that there is an n-line. Assume to the contrary that there is no n-line.

Let L be any line, and set $k_L = n+1-d$. If $\mathcal{M}$ is the set of lines parallel to L, then

$$(n^2-n-3+d)d = (v-k_L)d = \sum_{X\in\mathcal{M}} k_X = dn(n-1) - \sum_{X\in\mathcal{M}}(d_X-2),$$

since the number of lines parallel to L is dn. Because every line parallel to L has degree at most n-1 (i.e. $d_X \geq 2$ for all $X \in \mathcal{M}$), we conclude that $d \leq 3$. Hence every line has degree at least n-2. Furthermore, if $d=3$, then each of the 3n lines parallel to L has degree n-1, and if $d=2$, then two of the 2n lines parallel to L have degree n-2 and the other ones have degree n-1. Since this is true for every line, we obtain $3nb_{n-2} = 2b_{n-1}$, where b_k denotes the number of lines of degree k. We know two other equations, which are $n^2+n+1 = b = \sum b_k$ and $v(n+1) = \sum b_k \cdot k$

(this follows from equation (B_1) in section 1). In view of $b_k = 0$ for $k = n$ and $k \leq n-3$, we obtain $3b_{n-2} = n+3$; in particular, $b_{n-2} \leq n-2$. Hence $b_{n-2}(n-2) < v$ and it follows that L has a point p which is not contained in any line of degree n−2. This implies that p is contained in ½(n+3) lines of degree n−1 and in ½(n−1) lines of degree n+1 so that n is odd. Let L be any line of degree n−2. The inequality $b_{n-2} \leq n-2$ shows now that some point q of L is not contained in another line of degree n−2. Again we can determine the number t of lines of degree n−1 passing through q, and obtain this time $t = \frac{1}{2}n$. But n is not even, a contradiction.■

6.23 Corollary. Every (n+1,1)-design with $v \geq n^2-2$ points and n^2+n+1 lines, $n \geq 3$ and $n \neq 6$, can be embedded into a projective plane of order n.■

We conclude this section with a problem and a remark.

Given an integer $n \geq 2$, what is the smallest integer $v_0 = v_0(n)$ such that every (n+1,1)-design with $b \leq n^2+n+1$ lines and $v > v_0$ points can be embedded into a projective plane? We have proved in this section that $v_0 \leq n^2-\frac{1}{2}(\sqrt{5} - 1)n+O(\sqrt{n})$ but it seems likely that this bound is far from beeing best possible. Is it true that $v_0 \leq n^2-n+1$? If n is the order from a projective plane P, what kind of example can be constructed from P to obtain a lower bound for v_0?

Finally, we remark that results of embeddings of (n+1,1)-designs have sometimes applications in other fields of mathematics. For example, Füredi (1990) used such results in the characterization of quadrilateral free graph with the maximal possible number of edges.

7. An optimal bound for embedding linear spaces into projective planes

In this section we study linear spaces with $v \geq n^2-n+2$ points, $b = n^2+n+1$ lines, and a point of degree at least n+2. These spaces can not be embedded into a projective plane of order n, since each point of a projective plane of order n has degree n+1. It has been conjectured (de Witte, 1975b, and Erdős, Mullin, Sós, Stinson, 1985) that such structures do not exist, if n is large. However, this conjecture does not hold, since the closed complements of a Baer-subplane in a projective plane have the desired properties. We shall show that these are essentially the only examples for $v > n^2-\frac{1}{2}n+1$.

Suppose that L is a linear space which has a point q of degree at least n+2 and that L−q can be embedded into a projective plane P of order n. Then the lines passing through q form a parallel class Π of L−q. In order to determine L, we need information about the structure of Π in P. We shall obtain this information from the following two lemmata.

7.1 Lemma. Let P be a finite projective plane and denote by n the order of P. Suppose that q is a set of points and that Π is a set of lines with the properties that no line of Π is contained in q and that every point outside q lies on a unique line of Π. If $|q| \leq 2n-1$ and $|\Pi| < n+1+\sqrt{n}$, then the lines of Π are concurrent.

Proof. Let p be the set of points outside q, and set $v = |p|$. For every line L, we define $k_L = |L \cap p|$, and call k_L the degree of L (i.e., k_L is the degree of L in the space L = P−q). We have

$$(1) \qquad n^2-n+2 \leq n^2+n+1-|q| = |p| = v = \sum_{L \in \Pi} k_L.$$

If Π contains a line L of degree n+1, then $\Pi = \{L\}$, since every other line meets L in a point p, which is a point of p. If Π has a line N of degree n, then every line of Π has to contain the unique point of $N \cap q$. W.l.o.g. we may therefore assume that every line of Π has degree at most n−1.

Choose a line L of maximal degree of Π, put $d = n+1-k_L$ and $L \cap q = \{q_1,\dots,q_d\}$. Furthermore, denote by M_j the set of lines other than L of Π which contain q_j, and set $m_j = |\mathcal{M}_j|$, $j = 1,\ldots,d$. We may assume w.l.o.g. that $m_j \geq m_k$ for $j < k$. Finally, set $\mathcal{M} = \mathcal{M}_2 \cup \ldots \cup \mathcal{M}_d$, $m = |\mathcal{M}|$ and $a = |\mathbb{T}|-n-1$. Since every line of $\mathbb{T}$ has degree at most $n-1$, (1) implies that $a \geq 0$. Furthermore, our definitions yield

$$m_1+m = \sum_{j=1}^{d} m_j = |\mathbb{T}|-1 = n+a.$$

In order to prove Lemma 7.1, we have to show that $\mathcal{M} = \emptyset$. We shall do this in three steps.

Step 1. If $j,k \in \{1,\ldots,d\}$ with $j \neq k$, then every line of $\mathcal{M}_j$ has degree at most $n-m_k$.
For: This is true, since a line X of $\mathcal{M}_j$ intersects each of the lines of $\mathcal{M}_k \cup \{L\}$ in a point of q.

Step 2. $\mathcal{M}_1$ contains a line of degree $n-1$.
For: In view of $v > k_L+n(n-2)$, q_1 is contained in a line G with $G \neq L$ and $k_G \geq n-1$. If G is in $\mathcal{M}_1$, then G has degree $n-1$, since every line of $\mathbb{T}$ has degree at most $n-1$. It suffices therefore to show that G is a line of $\mathcal{M}_1$.

Assume to the contrary that $G \notin \mathcal{M}_1$. Then each of the points of $G \cap p$ lies on a line of $\mathcal{M}$. Consequently $m \geq k_G \geq n-1$. In view of $m_j \geq m_k$ for $j < k$, we obtain

$$m_1 \geq m_2 \geq \frac{m}{d-1} \geq \frac{n-1}{d-1}.$$

It follows that

$$n+a = m_1+m \geq \frac{n-1}{d-1} + n - 1,$$

i.e.

(2) $\quad n-1 \leq (a+1)(d-1).$

Using (1), we conclude that

$$\begin{aligned} n^2-(a+1)(d-1) \leq n^2-n+1 < v \leq |\mathbb{T}|k_L &= (n+1+a)(n+1-d) \\ &= n^2+n(a+2-d)-(a+1)(d-1). \end{aligned}$$

Hence $d < a+2$. This, our hypothesis $n > a^2$, and (2) imply $d = a+1$. Because of

$$a^2+a < n+a = \sum_{j=1}^{d} m_j \le dm_1,$$

it follows that $m_1 > a$. Now, $n+a = m_1+m$ and $m \ge n-1$ imply $m_1 = a+1$ and $m = n-1$. Since $k_X \le n-m_1$ for every line X of $\mathcal{M}$ (Step1) and $k_X \le k_L$ for every line X of $\mathcal{M}_1$ (because L is a line of maximal degree of $\mathbb{T}$), (1) shows that

$$n^2-n+2 \le v = \sum_{X\in\mathbb{T}} k_X = k_L + \sum_{X\in\mathcal{M}_1} k_X + \sum_{X\in\mathcal{M}} k_X$$
$$\le (m_1+1)k_L+m(n-m_1) = n^2+1-a(a+1).$$

We obtain $a(a+1) \le n-1$, and therefore

$$a(a+1) \le n-1 = m = \sum_{j=2}^{d} m_j \le (d-1)m_2 \le (d-1)m_1 = a(a+1).$$

Together this implies $m_1 = m_2 = a+1$, $a(a+1) = n-1$, $k_X = k_L = n-a$ for every line X of $\mathcal{M}_1$, and $k_X = n-m_1 = n-a-1$ for every line X of $\mathcal{M}$. This is a contradiction, since $k_X \le n-m_2$ for all $X \in \mathcal{M}_1$ by Step 1.

Step 3. $\mathcal{M} = \emptyset$.

For: By Step 2, $\mathcal{M}_1$ contains a line G of degree $n-1$. Since L is a line of maximal degree of $\mathbb{T}$, L has also degree $n-1$, i.e. $d = 2$. In particular, $\mathbb{T} = \{L\}\cup\mathcal{M}_1\cup\mathcal{M}_2$ so that $n+a = |\mathbb{T}|-1 = m_1+m_2$. Let q be the unique point other than q_1 of $G\cap q$. Since every line of $\mathcal{M}_2$ contains q (since it meets G in $q\cap(G-\{q_1\}) = \{q\}$), we have $m_2 \le 1$ and hence $m_1 \ge n+a-1 \ge n-1$. Step 1 shows that the line of $\mathcal{M}_2$, if there is one, has degree at most 1. Thus,

$$n^2-n+2 \le v = \sum_{X\in\mathbb{T}} k_X \le k_L+m_1(n-1)+m_2\cdot 1 \le m_1(n-1)+n.$$

This implies $m_1 \ge n$. Step 1 shows now that $k_X \le 0$ for every line X of $\mathcal{M}$. Since, by hypothesis, every line has degree at least 1, we obtain $\mathcal{M} = \emptyset$.

This completes the proof of Lemma 7.1.■

Remark. Let P be a projective plane of order $n = m^2$ which contains a Baer-subplane $(\mathcal{B},\mathcal{G})$. Let L be a line of $\mathcal{G}$, denote by q the set of points lying in $\mathcal{B}$ or on L, and set $\mathbb{T} = \mathcal{G}-\{L\}$. Then $|q| = 2n+1$ and $|\mathbb{T}| = n+m < n+1+\sqrt{n}$. Furthermore, every point outside of q lies on a unique line of $\mathbb{T}$ and no line of $\mathbb{T}$ is contained in q. This example shows that the bound for $|q|$ in Lemma 7.1 is almost best possible (is it best possible?).

7.2 Lemma. Let P be a finite projective plane and denote by n the order of P. Suppose that q is a set of points and that $\mathbb{T}$ is a set of lines with the properties that no line of $\mathbb{T}$ is contained in q and that every point outside q lies on a unique line of $\mathbb{T}$. Denote by B the set of points which lie on at least two lines of $\mathbb{T}$. If $2|q| < 3n+2$, then either B consists of just one point or $(B,\mathbb{T})$ is a Baer-subplane of P.

Proof. In view of Lemma 7.1, we may assume that $|\mathbb{T}| = n+1+m$ for a positive integer m with $m^2 \geq n$. We have to show that $(B,\mathbb{T})$ is a Baer-subplane of P. We shall do this in several steps. As in the last proof, we denote by p the set of points not in q, we set $v = |p|$, and we define the degree $k_L = |L \cap p|$ for every line L of P.

Step 1. Every line has degree at most n.
For: In view of $|\mathbb{T}| > n+1$, every line L has a point q which lies on at least two lines of $\mathbb{T}$. By definition, q is in q. This shows that L has degree at most n.

Step 2. If N is a line of degree n and if q is the point of $N \cap q$, then $N \notin \mathbb{T}$ and q is contained in exactly m+1 lines of $\mathbb{T}$.
For: Each of the n points of $N \cap p$ is contained in a unique line of $\mathbb{T}$. If N were in $\mathbb{T}$, then this would imply that every line of $\mathbb{T}$ passes through q, which is not possible, since $|\mathbb{T}| > n+1$. Consequently, $N \notin \mathbb{T}$. It follows that n of the lines of $\mathbb{T}$ meet N in a point of p and that q is contained in the remaining m+1 lines of $\mathbb{T}$.

Step 3. Every line of $\mathbb{T}$ contains at least m+1 points of B.
For: Let L be a line of $\mathbb{T}$, and denote by p a point of $L \cap p$. Then

$$n^2 - \frac{n}{2} < v = 1 + \sum_{p \in X} (k_X - 1).$$

Because every line has degree at most n, it follows that p lies on a line N of degree n. Step 2 shows $N \notin \mathbb{T}$. Let q be the point of $N \cap q$. Then each of the lines of $\mathbb{T}$ through q intersects L in a point of B. Therefore, Step 3 follows from Step 2.

Step 4. Every point of B is contained in exactly m+1 lines of Π.

For: Let q be a point of B, and denote by x the number of lines of Π through q. By the definition of B, $x \geq 2$.

Assume by way of contradiction that $x \neq m+1$. Then every line which contains q has degree at most n−1 (see Step 1 and Step 2). Since the x lines of Π which pass through q have degree at most n−m (Step 3), it follows that

$$n^2 - \frac{n}{2} < v \leq x(n-m)+(n+1-x)(n-1) = n^2-1-x(m-1).$$

In view of $n \leq m^2$, we obtain $2x \leq m+1$. Since $v > (n+1)(n-2)$, we see furthermore that q is contained in a line L of degree n−1. Step 3 and $m+1 \geq 1+\sqrt{n} > 2$ show that $L \notin \Pi$. Let $q' \neq q$ be the second point of $L \cap q$, and denote by y the number of lines of Π which contain q'. Obviously, $x+y = |\Pi|-k_L = m+2$. Consequently, $y \geq 2$, i.e. q' is a point of B, and $y = m+2-x \leq m$. In particular, $y \neq m+1$. As before, we can show that $2y \leq m+1$. This contradicts $2x \leq m+1$ and $x+y = m+2$.

Step 5. (B,Π) is a Baer-subplane of P.

For: Let L be a line of Π, and denote by x the number of points of B on L. By Step 4, $|\Pi| = 1+xm$, so $n+m = xm$. Since $x \geq m+1$ (Step 3) and in view of $m^2 \geq n$, we obtain $x = m+1$ and $n = m^2$. Consequently, every line of Π contains exactly m+1 points of B. This and Step 4 imply that any two distinct points of B are on a unique line of Π. Since any two lines of Π intersect by definition in a point of B, this proves that (B,Π) is a Baer-subplane of P.■

Now we start to study linear spaces with n^2+n+1 lines and a point of degree at least n+2. If a point has degree at most n, then Corollary 4.8 yields $v = n^2-n+2$. We exclude this case now, it will be studied in chapter 12.

Our first lemma still considers the general case $v \geq n^2-n+2$.

7.3 Lemma. Let L be a linear space with $v \geq n^2-n+2$ points and $b = n^2+n+1$ lines for some integer $n \geq 2$. Suppose every point has degree at least n+1 and some point has degree at least n+2. For every line L set $d_L = n+1-k_L$ and $t_L = \sum_{p \in L} (r_p-n-1)$. Then

a) Every line has degree at most n.

b) There exists a point of degree n+1. Every point of degree n+1 is contained in at least three lines of degree n.

c) If a line L contains a point of degree n+1, then $t_L \le d_L$.

d) For every line L, there exists a line N parallel to L such that $(d_L n - t_L)k_N \ge (v-k_L)d_L$.

e) Let H and L be different intersecting lines. If there is a line N of degree n which is parallel to H and L, then $d_H d_L \ge n$.

f) $t_N = 0$ for every line N of degree n.

<u>Proof</u>. a) For every line L we have $b \ge 1+k_L \cdot n$ with equality if and only if every point of L has degree n+1 and if L meets every other line, which means that every point outside of L has degree k_L. In view of $b = n^2+n+1$ and because there is a point of degree at least n+2, this implies that $k_L < n+1$.

b) Assume to the contrary that every point has degree at least n+2. Then $v(n+2) \le \sum r_p = \sum k_L$ shows that there is a line N of degree n. Since every point has degree at least n+2, N meets at least $k_N(n+1) = n^2+n$ other lines and every point outside N is contained in at least two lines parallel to N. This contradicts $b = n^2+n+1$.

Consequently, there is a point of degree n+1. Let p be any point of degree n+1. Since every line has degree at most n, we have $n^2-n+1 \le v-1 \le r_p(n-2)+c = n^2-n-2+c$, i.e. $c \ge 3$, for the number c of lines of degree n through p.

c) Let p be a point of degree n+1 contained in L. By part b), p is contained in a n-line N with $N \neq L$. Obviously, L intersects exactly $k_L-1+t_L = n-d_L+t_L$ lines which miss N. Since N is parallel to $b-1-k_N n-t_N \le n$ lines, this shows that $t_L \le d_L$.

d) L is parallel to $b-1-k_L n-t_L = d_L n-t_L$ lines. Consequently, if N is a line of maximal degree parallel to L, then

$$(v-k_L)d_L \le \sum_{p \notin L} (r_p-k_L) = \sum_{X \cap L=\emptyset} k_X \le (d_L n-t_L)k_N.$$

e) The Parallel-Lemma (see section 1) shows that $d_H d_L \ge n-t_N$. Since $t_N = 0$ (this will be shown in part f)), the assertion follows.

f) Let L be any line of degree n. Since every line has degree at most n (see Part a)), Part d) shows that $(n-t_L)n \ge v-k_L > n^2-2n$. Hence $t_L \le 1$, which implies that L contains a point of degree n+1. Part c) shows therefore that $t_L \le 1$.

Assume by way of contradiction that $t_L = 1$. Then L contains a unique point q of degree n+2. By part d), L is parallel to a line N of degree n. As before, $t_N \leq 1$. Let $H \neq L$ be the second line through q which is parallel to N. The Parallel-Lemma shows that $d_H d_L \geq n-t_N \geq n-1$. Because of $d_L = 1$ and $n+1-d_H = k_H \geq 2$, we conclude that $d_H = n-1$ and $t_N = 1$. Since N misses $b-1-k_N n-t_N = n-1$ lines, it follows (notice that $q \notin N$ and that $r_q-k_N = n+2-k_N = 2$)

$$v-n+1 \leq \sum_{p \notin N} (r_p-k_N) = \sum_{X \cap N=\emptyset} k_X \leq k_H+(n-2)n = n^2-2n+2.$$

But $v \geq n^2-n+2$, a contradiction.■

<u>7.4 Theorem</u> (Metsch, 1988b). Let L be a non-degenerate linear space with n^2+n+1 lines and $v > n^2-\frac{1}{2}n+1$ points. If some point has degree at least n+2, then $n = m^2$ is a perfect square and L can be embedded into the closed complement of a Baer-subplane in a projective plane of order n.

<u>Proof</u>. Since L is non-degenerate, we have $n \geq 2$ and $v \geq n^2-n+3$. By Corollary 4.8, every point has degree at least n+1. Thus, the hypotheses of Lemma 7.3 are satisfied and we can use its results. We prove Theorem 7.4 in several steps for which we use the following notation:

$s = n^2+1-v$,

$T = \{q \mid q$ is a point of degree at least $n+2\}$,

$v' = v-|T|$

$t = \sum_{q \in T} (r_q-n-1)$, and

$t_L = \sum_{p \in L} (r_p-n-1)$ for every line L,

Step 1. Every point of T has degree at least n+3. In particular, $2|T| \leq t$.

For: Assume to the contrary that some point q of T has degree n+2. In view of 7.3 f), q is not contained in a line of degree n. Since $v-1 > r_q(n-3)$, this shows that q lies on a line L of degree n−1. Because q has degree n+2, we have $t_L \geq 1$. Lemma 7.3 d) shows that L is parallel to a line N of degree n. If H denotes the second line through q which is parallel to N, then 7.3 e) shows that $2d_H \geq n$. We conclude that

$$v \le k_H+(r_q-1)(n-2) = k_H+n^2-n-2 \le n^2-\frac{n}{2}-1,$$

a contradiction.

Step 2. $|T| \le s-1$ and $v' > n^2-n+1$.

For: Let p be a point of degree n+1 (Lemma 7.3 b)), denote by e the number of lines of degree n which contain p, and set $f = n+1-e$. Furthermore, let $L_1,\dots,L_f$ be the lines which have degree at most n−1 and which pass through p, let $n+1-d_j$ denote the degree of L_j, and set $t_j = t_L$ for $L = L_j$. Then

$$n^2-s = v-1 = e(n-1)+\sum_{j=1}^{f}(n-d_j) \le e(n-1)+f(n-2) = n^2-n-2+e.$$

Hence, $e \ge n+2-s$, and

$$\sum_{j=1}^{f} d_j = e(n-1)+fn+s-n^2 = e(n-1)+(n+1-e)n+s-e = n+s-e \le 2s-2.$$

Together with Lemma 7.3 c) and f) it follows that

$$t = \sum_{q\in T}(r_q-n-1) = \sum_{j=1}^{f} t_j \le \sum_{j=1}^{f} d_j \le 2s-2.$$

Since $2|T| \le t$ (Step 1), we conclude that $|T| \le s-1$ and $v' = v-|T| > n^2-n+1$.

Step 3. Every point of T has degree at least n+4. In particular, $3|T| \le t$.

For: Assume to the contrary that there is a point $q \in T$ which has degree n+3 (see Step 1). We consider two cases.

Case 1. q is not contained in a line of degree n−1.

Since q is also not contained in a line of degree n (Lemma 7.3 f)) and in view of $v-1 > r_q(n-4)$, q lies on a line L of degree n−2. Let N be a line of degree n which is parallel to L (Lemma 7.3 d)) and denote by H and H' the two lines $\ne$ L through q which are parallel to N. Lemma 7.3 e) shows that $3d_H \ge n$ and $3d_{H'} \ge n$. Since every line through q has degree at most n−2, we obtain the following contradiction

$$n^2-\frac{n}{2} < v-1 < (k_H-1)+(k_{H'}-1)+(r_q-2)(n-3) = n^2-3-d_H-d_{H'}.$$

Case 2. q is contained in a line L of degree n−1.

As in Case 1, L is parallel to a line N of degree n, and q is contained in two lines H,H' ≠ L which are parallel to N. By Lemma 7.3e), $2d_H \geq n$ and $2d_{H'} \geq n$. It follows that $n^2-n < v-k_H \leq (r_p-1)(n-1)$ for every point p of H. Consequently, every point of H has degree at least n+2 and is therefore contained in *T*. Since the same holds for H', we obtain $k_H+k_{H'} \leq |T|+1$. On the other hand, since N is parallel to n lines and since every point outside of N is contained in at least one line parallel to N, we have

$$v-k_N \leq \sum_{X\cap N=\emptyset} k_X = k_H+k_{H'}+(n-2)n \leq |T|+1+n^2-2n.$$

But $v-|T| = v' > n^2-n+1$ (Step 2), a contradiction.

Step 4. Let p be a point of degree n+1, and denote by e the number of lines of degree n through p. Then $v'+e > n^2$.

For: Let f denote the number of lines of degree n−1 through p, set $g = n+1-e-f$, denote by $L_1,\dots,L_g$ the lines of degree at most n−2 through d, set $t_j = t_L$ for $L = L_j$, and denote by $n+1-d_j$ the degree of L_j. Then

$$n^2-s = v-1 = e(n-1)+f(n-2)+\sum_{j=1}^{g}(n-d_j)$$

Since $n-d_j \leq n-3$, we obtain

$$2e+f \geq n^2-s-(e+f+g)(n-3) = n^2-s-(n+1)(n-3) = 2n+3-s.$$

We obtain furthermore

$$\sum_{j=1}^{g} d_j = (e+f+g)n-e-2f+s-n^2 = n+s-e-2f.$$

Because no point of T lies on a line of degree n (Lemma 7.3 g)) or on a line of degree n−1 through p (Lemma 7.3 c) and Step 3), we have

$$t = \sum_{q\in T}(r_q-n-1) = \sum_{j=1}^{g} t_j.$$

In view of Lemma 7.3 d) it follows that

$$t \le \sum_{j=1}^{g} d_j = n+s-e-2f.$$

Since $3|T| \le t$ (Step 3), we conclude that

$$\begin{aligned} v'+e &= v-|T|+e \ge n^2+1-s-\frac{t}{3}+e \\ &\ge n^2+1-s-\frac{1}{3}(n+s-e-2f)+e \\ &= n^2+1-\frac{n}{3}-\frac{4}{3}s+\frac{2}{3}(2e+f) \\ &\ge n^2+1-\frac{n}{3}-\frac{4}{3}s+\frac{2}{3}(2n+3-s) \\ &= n^2+3+n-2s > n^2+3. \end{aligned}$$

This proves Step 4.

Step 5. If m denotes the unique integer with $(m-1)^2 < n \le m^2$, then every line L with $t_L \ge m$ has degree at most $n+1-m$.

For: If L contains a point of degree $n+1$, this follows from Lemma 7.3 d). If no point of L has degree $n+1$, then L is contained in T, so $k_L \le |T| \le s-1 \le n+1-m$.

Step 6. T contains at most one point q with $r_q \ge n+1+m$.

For: Assume to the contrary that T contains two points q_1 and q_2 of degree at least $n+1+m$. Let N be a line of degree n (such a line exists, since every point of degree $n+1$ is contained in a line of degree n). By Lemma 7.3 f), q_1 and q_2 are not points of N. Hence, N is parallel to at least $2m+1$ lines which contain q_1 or q_2. By Step 4, each of these lines has degree at most $n+1-m$. Because the number of lines L parallel to N is n, we conclude that

$$\begin{aligned} n^2+1-s-n+2m &= v-k_N+2m \le \sum_{p \notin N} (r_p-k_N) = \sum_{L \cap N=\emptyset} k_L \\ &\le (2m+1)(n+1-m)+(n-2m-1)n \\ &= n^2-2m^2+m+1 \le n^2-2n+m+1 \end{aligned}$$

This contradicts $2s < n$.

Step 7. L can be embedded into the closed complement of a Baer-subplane in a projective plane of order n.

For: The points of degree n+1 together with the lines not contained in T form an (n+1,1)-design L' with $v' > n^2-n+1$ points (Step 2) and at most $b = n^2+n+1$ lines. Furthermore, every line which has degree n in L has still degree n in L' (Lemma 7.3 f)). If p is a point of L', and if e the number of lines of degree n through p, then Step 4 shows that $v'+e \geq n^2$. Corollary 3.5 b) shows that L' can be embedded into a projective plane P of order n, if L' has n^2+n+1 lines, and the same follows from Lemma 3.6, if L' has at most n^2+n lines.

Let q be a point which has degree at least n+2. By definition, $q \in T$. Let Π be the set of lines of L' which contain q in L. Then Π is a parallel class of L'. Furthermore, every line of Π has degree at most n−2 in L', since a line of degree n of L can contains not the point $q \in T$ (Lemma 3.7 f)). Hence $v' \leq |\Pi|(n-2)$ and thus $|\Pi| \geq n+2$. Lemma 7.1 shows that $|\Pi| \geq n+1+m$. In the same way follows that each point of T has degree at least n+1+m. Therefore our last step implies $T = \{q\}$. Consequently, $v' = v-1 > n^2-½n$. If B denotes the set of points of P which lie in at least two lines of Π, then Lemma 7.2 shows that (B,Π) is a Baer-subplane of P. Thus, L' is embedded in P−B and any two distinct lines of Π are parallel in P−B. The linear space $L = L' \infty \Pi$ is therefore embedded in $(P-B) \infty \Pi$, which is the closed complement of (B,Π) in P.

This completes the proof of Theorem 7.4.■

In the following corollary, we summarize most of the results, we have proved up to now.

7.5 Corollary. Let L be a linear space with $v \geq n^2-n+2$ points and $b \leq n^2+n+1$ lines for some integer $n \geq 2$. Then one of the following cases occurs.

(1) L is a near-pencil.

(2) L can be embedded into a projective plane of order n.

(3) L can be embedded into the closed complement of a Baer-subplane in a projective plane of order n. In particular, n is a perfect square.

(4) $b = n^2+n+1$ and $v = n^2-n+2$. Furthermore, there is a point q of degree n which is contained in a line L of degree n+1 and n−1 lines of degree n. L meets every other line.

(5) $b = n^2+n+1$, $v \leq n^2-\frac{1}{2}n+1$, every point has degree at least n+1, and some point has degree at least n+2.

(6) $b = n^2+n+1$ and every point has degree n+1. If $n > 222$, then $v \leq n^2-\frac{1}{2}n+1$, if $n \leq 222$, then $v \leq n^2-\frac{1}{2}n+6$, and if $n \neq 6$, then $v \leq n^2-3$.

Proof. This follows from Corollary 5.7, Theorem 4.7, Corollaries 6.12, 6.14 and 6.23, and from Theorem 7.4.■

For every integer $n \geq 2$, we define A(n) to be the least integer ≥ 3 such that every non-degenerate linear space with $A(n) < v \leq b \leq n^2+n+1$ can be embedded into a projective plane of order n. Then A(n) is the "optimal bound" for embedding linear spaces into projective planes of order n. In general, the determination of A(n) seems to be very difficult, especially, if n−1 is not the order of a projective plane. If n−1 is the order of a projective plane, then the following lemma gives a very good lower bound for A(n).

7.6 Lemma. If n−1 is the order of a projective plane, then $A(n) \geq n^2-n+1$.

Proof. This follows from the fact that a projective plane of order n−1 can not be embedded into a projective plane of order n.■

For some values of n, we know the optimal bound A(n).

7.7 Corollary. Let n be a positive integer. If n is the order of a projective plane which contains a Baer-subplane, then $A(n) = n^2-\sqrt{n}+1$. In every other case, $A(n) \leq n^2-\frac{1}{2}n+6$.

Proof. If n is not the order of a projective plane with a Baer-subplane, then Theorem 7.5 shows that $A(n) \leq n^2-\frac{1}{2}n+6$.

Now suppose that n is the order of a projective plane P with a Baer-subplane B. Then the closed complement of B in P is a linear space with $n^2-\sqrt{n}+1$ points, which can not be embedded into any projective plane of order n. Because every non-degenerate linear space L with $n^2-\sqrt{n}+2 \leq v \leq b \leq n^2+n+1$ can be embedded into a projective plane of order n (Theorem 7.5), this yields $A(n) = n^2+1-\sqrt{n}$.■

7.8 Corollary. $A(p^{2s}) = p^{4s}-p^s+1$ for every prime p and every integer $s \geq 1$.

Proof. Set $n = p^{2s}$. Then there is a projective plane of order n which has a Baer-subplane (see for example Beutelspacher (1983), Satz 7.5). Corollary 7.8 follows therefore from Corollary 7.7.■

7.9 Corollary. $A(2) = 3$, $A(3) = 8$, and $A(4) = 15$.

Proof. Lemma 5.1 shows that $A(n) \leq n^2-1$ for all $n \geq 2$. This proves that $A(2) = 3$ and $A(3) = 8$, since the linear spaces E_1, E_2, E_3, and E_4 on 8 points can not be embedded into the projective plane of order 3. $A(4) = 15$ follows from Corollary 7.8.■

We already know another optimal bound which we want to recall here. For $n \geq 2$, let $A'(n)$ be the least integer ≥ 3 such that every non-degenerate linear space $A'(n) < v \leq b \leq n^2+n$ can be embedded into a projective plane of order n. Then

7.10 Corollary. $A'(3) = 8$ and $A'(n) \leq n^2-n+1$ for all $n \geq 3$. If n−1 is the order of a projective plane, or if n is the order of a projective plane containing a Baer-subplane, then $A'(n) = n^2-n+1$.

Proof Corollary 5.7 implies that $A'(3) = 8$ and $A'(n) \leq n^2-n+1$ for all $n \geq 3$. This proves that $A'(n) = n^2-n+1$, if n−1 is the order of a projective plane, since a projective plane of order n−1 can not be embedded into a projective plane of order n.

If n is the order of a projective plane with a Baer-subplane, then there is also an affine plane A of order n with a Baer-subplane B. The closed complement of B in A has n^2-n+1 points, n^2+n lines and can not be embedded into a projective plane of order n. This proves $A'(n) = n^2-n+1$ in this case.■

8. The Theorem of Totten

Up to now, we have studied linear spaces for which there is an integer n such that $(n-1)^2+(n-1)+1 = n^2-n+1 < v \le b \le n^2+n+1$. But there are more linear spaces with few lines. For example, what about linear spaces with n^2+n+1 points and n^2+n+2 or n^2+n+3 lines. Do they exist and, if yes, how do they look like? The answer was given by Totten (Totten, 1976c) who classified all restricted linear spaces, i.e. linear spaces satisfying $(b-v)^2 \le v$. If $n^2-n+2 \le v \le n^2+1$, this classification will not give us new information. However, $(b-v)^2 \le v$ includes the cases $n^2+2 \le v \le n^2+n+1$ and $b \le v+n$. In this section we give a proof for Totten's Theorem and we shall also classify all weakly restricted linear spaces.

The result of Stanton and Kalbfleisch (Lemma 2.4), already used for the proof of "$b \ge B(v)$", is very helpful again.

<u>8.1 Lemma</u>. Let L be a linear space and denote by n the unique integer with $n^2-n+2 \le v \le n^2+n+1$.

a) If $(b-v)^2 \le v$, then $b \le v+n$.

b) If $v < (b-v)^2 \le b$, then $v = 8$ and $b = 11$, or $v \in \{n^2+n, n^2+n+1\}$ and $b = v+n+1$.

c) If $(b-v)^2 \le b$ and if L has maximal line degree $k \ge n+2$, then L is a near-pencil, or one of the spaces $\mathbf{E_5}$, $\mathbf{E_6}$, and $\mathbf{E_7}$.

<u>Proof</u>. a) follows from $v < (n+1)^2$.

b) Set $b = v+n+z$. Then $(n+z)^2 = (b-v)^2 \le b = v+n+z \le n^2+2n+1+z$ shows that $z \le 1$ with equality only if $v \in \{n^2+n, n^2+n+1\}$ and $b = v+n+1$.

Now consider the case $z \le 0$. Then $(b-v)^2 > v$ yields $v \le n^2-1$ and $z = 0$. Hence, $b = v+n \le n^2+n-1$. Theorem 2.6 implies that $v = n^2-n+2$ and $b = n^2+n-1$. In view of $b = v+n$, we obtain $n = 3$, $v = 8$, and $b = 11$.

c) If $k = v-1$, then L is a near-pencil. We may therefore assume that $k \le v-2$. Lemma 2.4 shows that $b \ge f(v-2,v)$ or $b \ge f(n+2,v)$. In the first case, $v+n+1 \ge b \ge f(v-2,v) > 2v-5$, so $v \le n+5$. In the second case, $(v+n)(v-1) \ge (b-1)(v-1) \ge [f(n+2,v)-1](v-1) = (n+2)^2(v-n-2)$ and thus

$$0 \le n^2-v(n^2+3n+5)+n^3+6n^2+11n+8 = (n^2+n+1-v)(n+6-v)-v(n-2)-n^2+4n+2.$$

Now $v \geq n+6$ would imply that $(n+6)(n-2) \leq v(n-2) \leq 4n+2-n^2$, which holds only for $n = 2$. But if $n = 2$, then $v \leq n^2+n+1 = n+5$. Consequently, $v \leq n+5$ in every case. In view of $v \geq n^2-n+2$, we conclude that $n = 3$ and $v = 8$, or that $n = 2$.

Assume that $v = 8 = n+5$. Then $(b-v)^2 \leq b$ shows that $b \leq 11$. However, in view of $5 = n+2 \leq k \leq v-2 = 6$, we have $b \geq f(5,8) > 11$ or $b \geq f(5,8) > 11$, a contradiction.

Hence $n = 2$. Now we have $4 = n+2 \leq k \leq v-2 \leq n^2+n-1 = 6$. Thus, $v = 6$ and $k = 4$, or $v = 7$ and $k \in \{4,5\}$. Since $(b-v)^2 \leq b$, we obtain $b \leq v+3$. It is easy to see that L is Lin's Cross, if $v = k+2 = 6$, and that $L = \mathbf{E}_5$, if $v = k+2 = 7$.

Consider finally the case $v = 7$, $k = 4$ and $b \leq 10$, and let L be a line of degree 4. In view of $b \leq 10$, every line $\neq L$ has degree at most three. If b_x denotes the number of x-lines, then it follows that $9 \geq b-1 = b_2+b_3 \geq v-1 = 6$, and $42 = v(v-1) = \sum k_x(k_x-1) = 2b_2+6b_3+12$. Thus $b_2 = 6$ and $b_3 = 3$, or $b_2 = 3$ and $b_3 = 4$. Since there are only three points outside of L, $b_3 = 4$ is not possible. Hence, $b_2 = 6$ and $b_3 = 3$. It follows that $L = \mathbf{E}_7$. ■

<u>8.2 Lemma</u>. Suppose that L is a linear space with $v \leq n^2+n+1$ points and $b > n^2+n+1$ lines for some integer $n \geq 2$. If $b \leq v+n$ and if every line has degree at most $n+1$, then L is a complete projectively inflated affine plane.

<u>Proof</u>. In view of $v \geq n^2+2$, every point has degree at least $n+1$. We call the points of degree $n+1$ and the lines with a point of degree $n+1$ *real*. The other points and lines will be called *ideal*. Furthermore, we call a line *good*, if it has degree $n+1$ and a unique ideal point. Let v_o (or v') be the number of real (or ideal) points, and b_o (or b') be the number of real (or ideal) lines. Now we prove Lemma 8.2 in five steps.

Step 1. There exists an ideal point.

For: In view of $v \geq b-n \geq n^2+2$, every real point is contained in a line of degree $n+1$. In order to prove Step 1, we may therefore assume that it exists a line L of degree $n+1$. If every point outside L is real, then L meets every other line. In view of $b > n^2+n+1$, L contains an ideal point in this case.

Step 2. Every ideal point lies on a good line.

For: Let q be an ideal point, denote by $L_1,\ldots,L_r$, $r \geq n+2$, the lines through q, and define the numbers S_j as in Lemma 2.2. Since every point has degree at least $n+1$

and because every line has degree at most $n+1$, we have $S_j \leq 1$ for all indices j with equality if and only if L_j is a good line. In view of $b \leq v+n \leq v+r-2$, Lemma 2.2 c) shows therefore q is contained in a good line.

Step 3. Every real line contains at most one ideal point.

For: Assume to the contrary that there is a real line R which contains two ideal points q_1 and q_2. Let L_j be a good line containing q_j, $j = 1,2$. Then q_2 is contained in a line I parallel to L_1. Let q be a second point of I. Since I is parallel to L_1, q is an ideal point, hence it is contained in a good line L (Step 2). The Transfer-Lemma (Section 1) shows that q_2 lies on a line I' which meets L_1 and misses L. Since I' misses L, every point of I' has degree at least $n+2$. Hence, I' and every point of I' is ideal. Since L_1 is a good line, it has a unique ideal point, which is q_1. Hence, the point of intersection of L_1 and I' must be the ideal point q_1 of L_1. Now q_1 and q_2 lie on I' and R so that I' = R. But R is a real line and I' is ideal, a contradiction.

Step 4. $b_0 = n^2+n$, $v_0 \leq n^2$, and every ideal point lies on exactly n real lines. If $v_0 = n^2$, then every real line has exactly n real points.

For: Let q be any ideal point, L a good line through q, and p a real point of L. Since $v \geq n^2+2$, p is contained in a second line X of degree $n+1$. Let I be any line through q which is parallel to X, and let q' be a second ideal point of I. Since X is parallel to I, the point q' and the line I are ideal. Let L' be a good line passing through q' (Step 2). Then qp' is a real line for each of the n real points p' of L'. On the other hand, every real line through q has a point of degree $n+1$ and meets therefore L'. Consequently, q is contained in exactly n real lines. Since every point of $L-\{q\}$ is real and therefore contained only in real lines and because L meets every other real line, it follows that the number of real lines is $b_0 = n^2+n$. This implies that every real line has at most n real points (a line with $n+1$ real points would meet n^2+n other real lines).

Let x be any real point. Then each line through p has at most $n-1$ real points $\neq x$ and therefore the number of real points is $v_0 \leq 1+(n+1)(n-1) = n^2$. Furthermore, we have $v_0 = n^2$ if and only if every line through a real point has exactly n real points, i.e. if and only if every real line has exactly n real points.

Step 5. L is an inflated affine plane with a generalized projective plane at infinity.

For: In view of $b_0 = n^2+n$, there are $b-n^2-n \geq 2$ ideal lines. By definition, an ideal line contains only ideal points. Since any two ideal points are connected by an ideal line (Step 3), it follows that the ideal points and lines form a linear space D. It has v' points and b' lines. Theorem 2.3 shows that $b' \geq v'$. In view of $n^2+n+b' = b_0+b' = b \leq v+n = v_0+v'+n \leq n^2+n+v'$, we conclude that $b' = v'$ and $v_0 = n^2$. Step 4 shows now that every real line has exactly n real points. Consequently, the real lines induce an affine plane on the set of the real points and L is an affine plane of order n with D at infinity. Since $b' = v'$, D is a generalized projective plane.■

<u>8.3 Theorem</u> (Totten, 1976c). Every restricted linear space is one of the following structures:

(1) A near-pencil.

(2) A projective plane with at most n points deleted but not more than n−1 from the same line.

(3) an affine plane, or an affine plane with one infinite point, or a punctured affine plane with one infinite point.

(4) A complete projectively inflated affine plane.

(5) Lin's Cross.

<u>Proof</u>. Let L be a restricted linear space, and denote by n the unique positive integer with $n^2-n+2 \leq v \leq n^2+n+1$. If there is a line of degree at least n+2, then Lemma 8.1 c) shows that $L = E_3$, which is Lin's cross. If $b > n^2+n+1$ and if every line has degree at most n+1, then Theorem 8.3 follows from Lemma 8.2

We may therefore assume that $b \leq n^2+n+1$ and that every line has degree at most n+1. By Theorem 2.6, $b \geq B(v) \geq n^2+n-1$. In view of $(b-v)^2 \leq v$ this implies that $v \geq n^2$. Now Lemma 5.1 shows that L can be embedded into a projective plane of order n. It follows easily that (2) or (3) occurs.■

The original proof of the Theorem of Totten can be found in (Totten, 1975, 1976a, 1976b, 1976c) and (de Witte, 1975a). The main difference to our proof is that Totten did not use an algebraic approach (Lemma 2.2 played a crucial role in the proof of Lemma 8.2). This was first done by Fowler (1984) who gave a short proof for Totten's Theorem.

Definition. The *House of Nikolaus* is the affine plane of order two with one point at infinity.

8.4 Corollary. (Bridges, 1972). Every linear space L with $b = v+1$ is a punctured projective plane or the House of Nikolaus.

Proof. L satisfies $(b-v)^2 < v$ and is therefore restricted. If n denotes the integer with $n^2-n+2 \leq v \leq n^2+n+1$ then the Theorem of Totten shows that L can be embedded into a projective plane P of order n. Hence $v+1 = b \leq n^2+n+1$. In view of $b \geq B(v)$ (see Theorem 2.6), this yields $v = 5$ or $v = n^2+n$. If $v = n^2+n$, then L is a punctured projective plane, and if $v = 5 = b-1$, then L is the affine plane of order 2 with one infinite point.■

8.5 Corollary (de Witte, 1976). Every linear space L with two more lines than points is one of the following structures:

(1) A doubly punctured projective plane of order $n \geq 3$.

(2) The affine plane of order 2, which is also called the tetrahedron.

(3) Lin's Cross.

(4) The Fano quasi-plane, which is the affine plane of order 2 with the triangle at infinity.

(5) The affine plane of order 3 with one infinite point.

Proof. Since L is non-degenerate, it has at least four points and is therefore restricted. Corollary 8.5 follows now easily from the Theorem of Totten.■

The last two results have been proved earlier than the Theorem of Totten. After the classification of linear spaces with $b \leq v$ (de Bruijn and Erdös, 1948), it was quite natural to determine linear spaces with $b = v+1$ etc.. It is probably natural again to determine now linear spaces with $b \leq 1+\sqrt{(b-v)}$. However we are not going to do this, we just want to prove the following slight generalization of the Theorem of Totten.

<u>8.6 Theorem</u>. Every weakly restricted linear space is one of the following structures.

(1) A restricted linear space.

(2) An affine plane A of order n with a punctured projective plane on n or n+1 points at infinity.

(3) An complete projectively inflated punctured affine plane A.

(4) An affine plane of order 4 or 5 with the House of Nikolaus at infinity.

(5) The linear space obtained from the projective plane of order 3 by deleting two lines, their point of intersection, and two more points from each of these two lines.

(6) The linear space $\mathbf{E}_1$, $\mathbf{E}_6$, or $\mathbf{E}_7$.

<u>Proof</u>. Let n be the unique integer n with $n^2-n+2 \le v \le n^2+n+1$. If $v = 8$ and $b = 11$, then Lemma 5.2 and the remark following 5.2 show that $L = \mathbf{E}_1$ or that (5) occurs. If $(b-v)^2 \le v$, then L is restricted so that (1) occurs, and if some line has degree at least n+2, then Lemma 8.1 c) shows that L is a near-pencil or $\mathbf{E}_3$, or $\mathbf{E}_6$, or $\mathbf{E}_7$ (the linear space $\mathbf{E}_3$, which is Lin's Cross, and the near-pencils are restricted). We may therefore assume that $v < (b-v)^2 \le b$ and that every line has degree at most n+1. Lemma 8.1 b) shows that $v \in \{n^2+n, n^2+n+1\}$ and $b = v+n+1$ in this case. We have to show that L is one of the structures described in (2), (3), and (4).

Since every line has degree at most n+1, every point has degree at least n+1. We call a point of degree n+1 *real*, and a line with a real point also *real*. The other points and lines are called *ideal*. Furthermore, we call the lines of degree n+1 *long*. If $v = n^2+n+1$, then a real point is contained in n+1 long lines. If $v = n^2+n$, then every real point is contained in a unique n-line and in n long lines. The rest of the proof will be done in several steps.

Step 1. *Every long line is real, so every real line meets every long line. If $v = n^2+n$, then any two real n-lines are parallel. There exists a real point.*

For: Assume that there is an ideal long line L. Then L meets at least $(n+1)^2$ other lines. In view of $b \le n^2+2n+2$, it follows that $b = n^2+2n+2$ and that L meets every other line. In particular, every point outside L has degree $k_L = n+1$. Let p be a point outside of L. Then p has degree n+1 and is therefore contained in a line G of degree n+1, which meets L. Let q be a point on L which is not the point of intersection of L and G. Since q is ideal, it lies on a line I parallel to G. Since G has degree n+1, every point of I has degree at least n+2. But every point outside of L has degree n+1, a contradiction. Consequently, every long line is real.

Now suppose that $v = n^2+n$ and that R and N are n-lines which intersect in a point q. We have to show that R or N is ideal. W.l.o.g. we may assume that R is real. Let p be a real point of R. Then $p \neq q$, since q is ideal (a real point lies on a unique line of degree n). The point p lies on n long lines and one of these long lines misses N. Hence, every point of N is ideal. By definition, N is ideal.

If every point were ideal, then every line would have degree at most n and we would have $v(n+2) \leq bn$. But $v(n+2) > bn$, so there are real points.

Step 2. *If $n = 2$, then L is the punctured affine plane of order 2 with the triangle at infinity.*

For: If $n = 2$, then every line has degree 2 or 3. Thus, if b_k is the number of k-lines, then $v(v-1) = 2b_2+6b_3$ and $v+3 = b = b_2+b_3$. In view of $v \in \{6,7\}$, we obtain $v = 6$, $b_3 = 3$, and $b_2 = 6$. Since $v = 6$, the three 3-lines can not be pass through the same point. Hence, the 3-lines form a triangle, so L is the 'Fano quasi-plane with its midpoint deleted'. This is the punctured affine plane of order 2 with the triangle at infinity (case (3) of the theorem).

Step 3. *If there is an ideal point q such that every point $\neq q$ lies on at least n real lines, then* L *is one of the structures described in (2), (3), and (4).*

For: Since every real point lies on at least n long lines, there is a long line L not containing q. Since every point of L lies on at least n real lines, it follows that there are at least n^2 real lines.

First consider the case $v = n^2+n+1$, and let $D = (p, L_r)$ be the partial linear space consisting of all points and the real lines. Since every real line has degree n+1 in this case, any two lines of D intersect. Consequently, D is dual to an (n+1,1)-design with $|L_r| \geq n^2$ points and $|p| = n^2+n+1$ lines. Lemma 3.5 a) shows therefore that D can be extended to a projective plane $P = (p, L_r \cup B)$ of order n. The ideal points of L are the points of the lines of *B*.

Let B be a line of *B*. We shall show that the ideal lines which are contained in B induce a linear space L(B) on B. For this we have to show that B contains at least two ideal lines, and that every line q_1q_2 of L with $q_1,q_2 \in B$ is ideal and contained in B. Let q_1 and q_2 be two points of B. W.l.o.g. $q_2 \neq q$. By hypothesis, q_2 lies on n lines of $\mathcal{L}_r$. These n real lines cover every point which is not in B. Thus, every other line through q_2 is ideal and contained in B. Hence, q_1q_2 is ideal and contained in B. Since every ideal point is contained in at least two ideal lines, B contains at least two ideal lines. Hence, L(B) is a linear space.

Since every linear space has at least as many lines as points, it follows that B contains at least n+1 ideal lines and the same holds for every line of $\mathcal{B}$. Thus, L has at least $|\mathcal{B}|(n+1)$ ideal lines. Because there are $n^2+n+1-|\mathcal{B}|$ real lines and in view of $b = n^2+2n+2$, we conclude that $|\mathcal{B}| = 1$ ($|\mathcal{B}| = 0$ is not possible). Consequently, there is a unique line B in $\mathcal{B}$ and L is an affine plane of order n with L(B) at infinity. In view of $b = n^2+2n+2$, L(B) has n+2 lines. Because L(B) has n+1 points, Corollary 8.4 shows that L(B) is a punctured projective plane or the House of Nikolaus. In the first case (2) occurs. In the second case, the number n+1 of points of L(B) equals to 5, so n = 4 and (4) is fulfilled.

Now suppose that $v = n^2+n$. Then there are real n-lines, which are mutually parallel (Step 1). Let $D' = (p \cup \{\infty\}, \mathcal{L}_r)$ be the structure consisting of the points, the real lines, and a new point ∞, which lies on every real n-line. Then every line of D' has degree n+1, and any two lines of D' intersect. As before, it follows from Lemma 3.5 a) that D' can be extended to a projective plane $P = (p \cup \{\infty\}, \mathcal{L}_r \cup \mathcal{B})$.

This time, the lines B of $\mathcal{B}$ may contain the point ∞. But as before we can show that the ideal lines contained in B induce a linear space L(B) on the points of $B-\{\infty\}$. If $\infty \notin B$, then L has n+1 points and at least n+1 lines, and if $\infty \in B$, then L(B) has n points and at least n lines. Hence, there are at least $|\mathcal{B}|n$ ideal lines. Since there are $n^2+n+1-|\mathcal{B}|$ real lines and in view of $b = n^2+2n+1$, we conclude that $|\mathcal{B}|(n-1) \leq n$. Since we can exclude the case n = 2 (Step 2), it follows that $\mathcal{B}$ consists of a unique block B. There are two possibilities now.

The first one is that ∞ is not a point of B. Then L(B) has n+1 points. Since there are n^2+n real lines and in view of $b = n^2+2n+1$, L(B) has n+1 lines. Hence, L(B) is a generalized projective plane and L is a punctured affine plane (induced by the real lines on the n^2-1 real points) with L(B) at infinity.

Finally suppose that ∞ is a point of B. Then L is an affine plane with L(B) at infinity. This time L(B) has n points and n+1 lines. Corollary 8.4 shows that L(B) is a punctured projective plane or the House of Nikolaus. In the second case, we have $n = 5$, since the House of Nikolaus has 5 points. This proves Step 3.

Step 4. *Suppose there exists an ideal point q which has degree at least $n+3$. Then every point other than q is contained in at least n real lines.*
For: As in Step 2 of the proof of Lemma 8.2, it follows from Lemma 2.2 that q lies on a line L of degree $n+1$ which has n real points. These real points of L are contained only in real lines. Furthermore every point outside of L is joined to each of these points by a real line. Hence, every point other than q lies on at least n real lines.

Step 5. *Suppose that $n \neq 2$ and that every ideal point has degree $n+2$. Then every point is contained in at least n real lines.*
For: Assume to the contrary that there is an ideal point q which lies on three ideal lines H, I and J. We may assume that $|I| \geq |H| \geq |J|$. We have

$$v-1 \leq (n-1)n+(k_H-1)+(k_I-1)+(k_J-1).$$

Consider first the case that H meets every long line. Let p be a real point. Since H has degree at most n and since p lies on at least n long lines, it follows that H has degree n and that p lies only on n long lines, which means that $v = n^2+n$ and that the n-line through p misses H. Since every ideal point has degree $n+2$, the line H meets n^2+n other lines, and I meets $(k_I-1)2 \geq (k_H-1)2 = 2n-2$ lines which miss H. Hence there are at least $1+n(n+1)+2(n-1)$ lines. This is a contradiction, since $b = v+n+1 = n^2+2n+1$ and $n \neq 2$.
Now suppose that H misses a line L of degree $n+1$ and denote by m be the number of lines parallel to L. Since H is the unique line through q which misses L (ideal points have degree $n+2$), the line I meets L. Since every point of I is ideal, it follows that every point $\neq L \cap I$ of I lies on a line missing L. Hence $m \geq k_I-1$. By Step 1, the line L contains a real point p. Since p lies on at least n long lines, it lies on a long line G which meets H. By the Transfer-Lemma (Section 1), every point $\neq H \cap G$ of H lies on a line which meets L and misses G. Since every idal point of L has degree $n+2$ and lies on a unique line missing G, it follows that L has at least k_H-1 ideal points. This implies that L meets at least n^2+n+k_H-1 other lines so that $b-1 \geq n^2+n+k_H-1+m \geq n^2+n+k_H-1+k_I-1$. Using the above upper bound for

the number of points, we see that $n+1 = b-v \geq 2n+1-k_J$. Since $k_J \leq k_H \leq k_I$ and because ideal lines have degree at most n (Step 1), it follows that $k_J = k_H = k_I = n$ and we obtain equality in the above inequalities; in particular $v = n^2+3n-1$. In view of $b \leq n^2+2n+2$, we obtain $n = 3$, $b = n^2+2n+2 = 17$ and $v = n^2+n+1 = 13$. We have furthermore shown that L contains exactly $k_H-1 = 2$ ideal points.

Let b_k be the number of k-lines, $k = 2,3,4$. Then $17 = b = b_2+b_3+b_4$ and $142 = v(v-1) = 2b_2+6b_3+12b_4$. Since each of the seven points on H, I, or J has degree at least 5 and since the other six points have degree at least 4, we have furthermore $2b_2+3b_3+4b_4 = \sum k_L = \sum r_p \geq 7\cdot5+6\cdot4 = 59$. Hence, $2b_3+5b_4 = 54$ and $b_3+2b_4 \geq 25$. It follows that $b_4 \leq 4$. But each of the two real points of L is contained in four lines of degree 4, a contradiction.

Step 2, 3, 4, and 5 complete the proof of Theorem 8.6.■

Corollary 8.7. (Totten, 1976d) Every linear space with three more lines than points is one of the following structures:

(1) A projective plane of order $n \geq 3$ with 3 points deleted, and these three points are not collinear if $n = 3$.

(2) The affine plane of order 3, the affine plane of order 4 with one point at infinity, or the punctured affine plane of order 3 with one infinite point.

(3) The projective plane of order 3 with a near-pencil on four points at infinity.

(4) The linear space obtained from the projective plane of order 3 by deleting two lines, their point of intersection, and two more points from each of these two lines.

(5) The space obtained from the Fano quasi-plane by removing a point of degree 3.

(6) The linear space with 5 points which has parallel lines of degree 2 and 3.

(7) $\mathbf{E}_1$, $\mathbf{E}_6$, or $\mathbf{E}_7$.

Proof. Clearly, $v \geq 4$. Since the only linear spaces on 4 points are the affine plane of order 4 and the near-pencil on four points, we have $v \neq 4$. If $v = 5$, then it is easy to see that (6) occurs.

Now suppose that $v \geq 6$ so that $(b-v)^2 = 9 \leq b$. Then one of the cases of Theorem 8.3 or Theorem 8.6 occurs. Using $b = v+3$, we see that 8.3 (1) and 8.3 (5) are not possible, that 8.3 (2) implies (1), 8.3 (3) implies (2), and 8.3 (4) implies (3) of our corollary. 8.6 (2) and 8.6 (4) are also not possible. 8.6 (3) corresponds to (5), 8.6 (5) to (4), and 8.6 (7) to (7) of the corollary.■

Corollary 8.8. Suppose L is a weakly restricted linear space which is not restricted. Then L is one of the following structures.

(1) A punctured restricted linear space.

(2) A complete inflated affine plane with a punctured projective plane at infinity.

(3) $\mathbf{E}_1$, $\mathbf{E}_6$, $\mathbf{E}_7$, or an inflated affine plane of order 4 or 5 with the House of Nikolaus at infinity.■

A consequence of the Theorem of Totten is that every linear space with $n^2-n+2 \leq v \leq n^2+n+1$ and $b \leq v+\sqrt{v}$, $n >> 1$, is a subspace of a projective plane of order n, or an inflated affine plane of order n. Theorem 8.6 shows that this bound for b is not best possible. But how much can it be improved? In general, the bound $b \leq v+{}^2\sqrt{v}+{}^4\sqrt{v}+1$ is too weak. The closed complements of a Baer-subplane in a projective plane are counterexamples. Is it true that the condition $b \leq v+{}^2\sqrt{v}+{}^4\sqrt{v}$ is strong enough? The following lemma, which has been proved by Erdös, Fowler, Sós and Stinson (1985) for $s = 0$, may be helpful to answer this question.

8.9 Lemma. Let $\mathbf{L} = (p, L)$ be a linear space and denote its number of points by $n^2+n+1-s$. Suppose that L_r is a maximal subset of L such that the partial linear space (p, L_r) can be embedded into a projective plane P of order n and put $|L_r| = n^2+n+1-z$. If $z \geq 1$, $s+z \leq n$, i.e. $n^2+1+s \leq |L_r| = n^2+n$, then the linear space L has at least $n^2+n+1+z(n-s)-(z-1)^2$ lines.

Proof. We call the lines of L_r *real* lines and the other lines of L *ideal* lines. Let Z be the set consisting of the z lines of P which do not belong to L. If X is any line of Z, then we consider X as a set of points of L and we want to show the following: There are $n+1-s$ ideal lines which meet X in at least two points, and at least $n+2-s-z$ of these lines are contained in X. Then, if X is a line of Z, there are $n+1-s$ ideal lines which meet X in at least two points, and for every other line X' of Z there are at least $n+2-z-s$ ideal lines which are contained in X'. Hence there are at least $n+1-s+(z-1)(n+2-s-z)$ ideal lines so that L has at least $|\mathcal{L}_r|$+n+1−s+(z−1)(n+2−s−z) lines, which proves the lemma. Since the structure $(p,\mathcal{L}_r)$ is embedded in **P**, we have the following.

(P) Any two points of L of the same line X of Z are joined by an ideal line. If x is a point of L lying on a unique line X of Z, then every ideal line through x is contained in X.

Let X be any line of Z. Then X, considered as a set of points of L, has at least n+1−s ≥ 2 points, since $v = n^2+n+1-s$. Let X' be the subset of X consisting of those points which lie in no other line of Z. In view of $|Z| = z$, we have $|X'| \geq |X|-(z-1) \geq n+2-z-s \geq 2$.

No line of L contains all points of X. (Assume to the contrary that $X \subseteq L$ for a line L of L. Then (P) shows that L is ideal and that L = X. But then $(p,\{L\}\cup\mathcal{L}_r)$ is still embedded in **P**, contradicting the maximality of $\mathcal{L}_r$.) Hence if $\mathcal{L}_X$ is the set of the ideal lines which meet X in at least two points, then $|\mathcal{L}_X| \geq 2$, so that the lines of $\mathcal{L}_X$ induces a linear space L_X on X. The Theroem of De Bruijn and Erdős shows therefore that $|\mathcal{L}_X| \geq |X| \geq n+1-s$. If there is an ideal line containing every point of X', then every point of X' lies on a second line of the linear space L_X. If there is no ideal line containing all points of X', then the ideal lines which have at least two points in X' induce a linear space on X', and the Theorem of De Bruijn and Erdős shows that there are at least $|X'|$ such lines. Hence, in either case, there are at least $|X'| \geq n+2-s-z$ ideal lines which meet X' and have at least two points in X. By (P), these lines are contained in X. This completes the proof of the lemma.■

9. Linear spaces with n^2+n+1 points.

A consequence of the Theorem of Totten is that the inflated affine planes are the only linear space satisfying $v \le n^2+n+1 < b$ and $b \le v+n$ for some integer $n \ge 3$. It seems very likely that the bound $b \le v+n$ can be improved. However the problem seems to be very difficult and only the cases that L has either the maximal number n^2+n+1 of points or the minimal number n^2+n+2 of lines have been studied. It is the purpose of this section to study the case $v = n^2+n+1$. The case $b = n^2+n+2$ will be handled in section 11.

Linear spaces L with $v = n^2+n+1$ points and $b > v+n$ lines were first studied by Erdös, Fowler, Sós and Wilson (1985). They proved that L is an inflated affine plane, if $b \le n^2+2.14 \cdot n$. Then Blokhuis, Schmidt and Wilbrink (1988) showed even that it suffices to demand $b \le n^2+3n-4\sqrt{n}$. The coefficient 3 of the linear term is best possible in general. In fact, if n+1 is the order of a projective plane P, then P contains subspaces with n^2+n+1 points and n^2+3n+1 lines. Another example with n^2+3n+1 lines can be obtained from a projective plane of order n, if we replace two lines by a near-pencil. It seems likely that every linear space with $v = n^2+n+1 < b \le n^2+3n$ is an inflated affine plane. It would already be nice to prove that there is an absolute constant c such that any linear space satisfying $v = n^2+n+1 < b \le n^2+3n-c$ is an inflated affine plane. We shall prove that the bound $b \le n^2+3n+1-2\sqrt{(2n)}$ is strong enough for large n. The proof uses essentially the same techniques as the proof of Blokhuis, Schmidt and Wilbrink (1988).

<u>9.1 Theorem</u>. Suppose that L is a linear space with n^2+n+1 points and denote its number of lines by b. If L is neither a generalized projective plane nor an inflated affine plane then $b \ge n^2+3n+1-2\sqrt{(2n)}$ or $b \ge n^2+3n-143$.

The proof will be obtained after several lemmas. Throughout in this chapter, L denotes a linear space with n^2+n+1 points which is not a generalized projective plane or an inflated affine plane. We shall assume that L has less than n^2+3n-1 lines.

9.2 Lemma. Every point has degree at least n+1 and every line has degree at most n+1. There exists three non-collinear points of degree n+1.

Proof. We want to apply the Lemma of Stanton and Kalbfleisch (Lemma 2.4). Since $b < f(n+2,v)$ and $b < f(v-2,v)$ and because L has no line of degree v−1, this lemma shows that every line has degree at most n+1. It follows that every point has degree at least n+1 with equality if and only if it is contained only in lines of degree n+1.

Now we claim that there exists a line of degree n+1. Assume to the contrary that every line has degree at most n. Then $v(v-1) \leq bn(n-1)$, which implies that $b > n^2+3n+1$, since $v = n^2+n+1$. This contradiction shows that there is a line L of degree n+1.

Now assume that every point has degree at least n+2. Then L meets at least $(n+1)^2$ other lines. Denote by m the number of lines missing L. Since each of the n^2 points outside of L lies on at least one line missing L and because every line (missing L) has degree at most n+1, we have $n^2 \leq m(n+1)$, which implies $m \geq n$. Consequently, $b \geq 1+(n+1)^2+m \geq n^2+3n+2$. This contradiction shows that there exists a point p of degree n+1.

It remains to show that there are three non-collinear points of degree n+1. Assume to the contrary that there is a line H which contains all points of degree n+1. Then p is on H. Let L_1 and L_2 be two lines $\neq H$ through p. Then every point $\neq p$ of L_1 or L_2 has degree at least n+2. Hence, L_1 meets at least n^2+2n other lines and every point of $L_2-\{p\}$ lies on a line which misses L_1. It follows that $b \geq 1+n^2+2n+|L_2-\{p\}| = n^2+3n+1$, a contradiction. Hence there are three non-collinear points of degree n+1.■

A point of degree n+1 will be called real. A line will be called *long*, if it has degree n+1, and otherwise *short*. Thus, a point is real if and only if it lies only on long lines. For every point p, we set $e_p := r_p-n+1$, and for every line L we set $e_L = \Sigma e_p$ where the sum runs over all points p of L. The values e_p and e_L are called the *excess* of p and L, respectively. The following lemma is immediate.

9.3 Lemma. a) A long line L meets n^2+n+e_L other lines.
b) Suppose that H and L are long lines which meet in a point p. Then there are $n^2+n+1+e_H+e_L-e_p$ lines which meet H or L.■

9.4 Lemma. a) Suppose that G, L, and M are three concurrent long lines, and that there is a line H meeting exactly one of them. Then $b \geq n^2+n+1+2(k_H-1)$.
b) If a line H has degree at most n−1, then $b \geq n^2+n+1+2(k_H-1)$.

Proof. a) Denote the point of intersection of G, L, and M by p. We may assume that H meets M and misses G and L. Since e_G-e_p is the number of lines which meet G and miss M, the Transfer-Lemma (section 1) shows that $e_G-e_p \geq k_H-1$. Similarly, $e_L-e_p \geq k_H-1$. The assertion follows now from the preceding lemma.

b) Let p be a real point. Since H has degree at most n−1, the point p lies on two lines G and L which miss H. Since p lies also on a line which meets H and because every line through p is long, Part b) follows from Part a).■

9.5 Lemma. There exists at most one line of degree n.

Proof. In this proof we call a long line *real*, if it contains a real point. Clearly, distinct real lines meet.

Suppose that N is an n-line. Then every real point lies on a unique (real) line missing H. We want to show that there is a point s such that s lies on every real line which misses N. Assume to the contrary there are real lines L_1, L_2, and L_3 such that the point $x := L_1 \cap L_2$ is not on L_3, and let p be a real point of L_3. Then the line xp meets N, since L_3 is the unique line through p missing N. Since L_3 is a real line, Lemma 9.4 a) shows that $b \geq n^2+n+1+2(k_N-1) = n^2+3n-1$, which is a contradiction. Hence, there exists a point s lying on every real line which misses H.

Assume by way of contradiction that there is a second n-line N'. Then there is also a point s' lying on every real line missing N'. The lines N and N' are disjoint (if N and N' were not disjoint, then, since every point of N or N' has degree at least n+2, there would be at least $1+k_N(n+1)+2(k_{N'}-1) = n^2+3n-1$ lines meeting N or N'). Still following the paper of Blokhuis, Schmidt, Wilbrink (1988), we consider two cases.

Case 1. The points s and s' are distinct.
Denote the degree of s and s' by n+1+e resp. n+1+e', let p be a real point which is not on the line through s and s' (Lemma 9.2), and consider the real line L through p and s. If x is a real point of L, then $x \notin N'$ (since real points lie only

on long lines) so that x lies on a unique (real) line missing N'. Since s' lies on all real lines missing N', we see that xs' is the line through x missing N'. Since s' lies on e'+1 lines which miss N', it follows that the line L has at most e'+1 real points. Hence L' has at least n+1−e' non-real points. Since the point s has excess e and lies on L, it follows that the excess e_L of L is at least e+(n−e'). Similarly, the line L' through p and s' has excess at least e'+(n−e). Lemma 9.2 implies therefore that $b \geq n^2+n+1+e_L+e_{L'} \geq n^2+3n+1$, a contradiction.

Case 2. s = s'.

We denote by f the number of lines through s which miss N and N'. Let $p \neq s$ be a real point and let M be a real line through p with $s \notin M$. Then the line xs is parallel to N and N' for every real point x of M. Hence M has at most f real points, i.e. $e_M \geq n+1-f$. Now consider the line L := ps, which misses N and N'. Since N and N' are disjoint, the number of lines meeting L and $N \cup N'$ is $k_L(k_N+k_{N'})-k_Nk_{N'} = n^2+2n$. Since L meets f lines missing N and N', it follows that L meets at least $n^2+2n+f-1$ other lines. Hence $e_L \geq n+f-1$ so that Lemma 9.3 shows that $b \geq n^2+n+1+e_L+e_M \geq n^2+3n+1$. This contradiction completes the proof of the lemma.■

<u>9.6 Lemma</u>. If v' be the number of real points and if b' be the number of long lines which contain a real point, then $b'(v'+n) \geq v'(n+1)^2$, $b' \leq n^2$, and $2v' \leq n^2$.

<u>Proof</u>. For $j \geq 1$, let b_j be the number of long lines with j real points. Then

$$\sum_{j\geq 1} b_j = b'.$$

Counting in two ways the number of incident pairs (p,L) of real points and long lines, and the number of triples (p,q,L) of real points p and q and long lines L with $p,q \in L$, we get

$$\sum_{j\geq 1} b_j \cdot j = v'(n+1) \quad \text{and} \quad \sum_{j\geq 1} b_j \cdot j(j-1) = v'(v'-1).$$

Now we use the Cauchy-Schwarz inequality $(\Sigma b_j)(\Sigma b_j \cdot j^2) \geq (\Sigma b_j \cdot j)^2$ to obtain $b'v'(v'+n) \geq [v'(n+1)]^2$, i.e. $b'(v'+n) \geq v'(n+1)^2$. If we can show that $b' \leq n^2$, then this implies that $2v' \leq n^2$.

Assume by way of contradiction that $b' > n^2$ and let L' be the set of the b' long lines which have at least one real point. Then any two lines of L' meet. Let L_r be any maximal set of mutually intersecting lines satisfying $L' \subseteq L_r$. Then every line of L_r is long, since it meets all lines passing through a real point. Hence, the partial linear space consisting of the n^2+n+1 points of L and the lines of L_r is dual to an (n+1,1)-design and can be extended to a projective plane of order n (use the result dual to Corollary 3.5 b) and the assumption that $b' > n^2$). Since L is not a projective plane or an inflated affine plane of order n, we have $|L_r| = n^2+n+1-z$ for some integer $z \geq 2$. In view of $b < n^2+3n-1$, Lemma 8.9 shows that $z > n$. But now $b' \leq |L_r| \leq n^2+n+1-z \leq n^2$, a contradiction.■

9.7 Lemma. Suppose that q is a point of degree n+2, let H be a short line of maximal degree through q, set $k_H = n+1-d$, and denote by s+1 the number of short lines through q.

a) We have $(d+1)s \leq n$.

b) If p is a point $\neq q$ of H and if n+1+x is the degree of p, then p lies on $d-(s-1)x$ short lines.

c) If H has at most two points of degree n+2, then H meets a short line G satisfying $n-71 \leq k_G \leq n-1$.

d) If q' is a second point of degree n+2 on H and if H is also a short line of maximal degree through q', then H has degree at least $n+1-2\sqrt{(2n)}$.

Proof. Let $H_1,\dots,H_s$ be the short lines $\neq$ H through q and denote the degree of H_j by $n+1-d_j$. If we count the number of points using the lines through q, we see that $d+\Sigma d_j = n$.

a) By hypothesis, we have $d \leq d_j$ for all j. Therefore $d+\Sigma d_j = n$ implies that $d(s+1) \leq n$.

b) The point p lies on $x+d_j$ lines which miss the line H_j. It follows that p lies on at most $\Sigma(x+d_j) = sx+\Sigma d_j = sx+n-d$ lines which miss one of the lines H_j. Thus p lies on at least $r_p-(sx+n-d) = d+1-(s-1)x$ lines which meet every line H_j. In order to prove part b), it suffices therefore to show that p lies on at most one long line which meets every short line H_j.

Assume to the contrary that p lies on two long lines L and M which meet every line passing through q. Then the line through q which is parallel to L (or

M) is long. Therefore every point of $L \cup M$ has degree at least n+2. It follows that there are at least n^2+3n+1 lines which meet L or M, which is a contradiction.

c) Suppose that H has at most two points of degree n+2. We want to show first that this implies that $6d > n$. To see this, let p be a real point and let L be a line through p which meets H. Then H meets at least $2(k_H-1)-2 = 2(n-d-1)$ lines which are parallel to L. If x is the number of non-real points on L, then L meets at least n^2+n+x other lines. It follows that $n^2+n+1+x+2(n-1-d) \le b$. Since $b \le n^2+3n-2$, we obtain $x \le 2d$. Hence, L has at least $n+1-2d$ real points. Since L was any line through p which meets H, it follows that the number of real points is at least $k_H(n-2d) > n^2-3dn$. Lemma 9.6 shows therefore that $6d > n$.

In view of $(s+1)d \le n$, we conclude that $s \le 4$. On the other hand, we have $s \ge 1$ (the point q is not real and lies therefore on at most n long lines) so that $2d \le n$, i.e. $k_H \ge 1+n/2$.

We have to show that H meets a line G with $n-71 \le k_G \le n-1$. If $n \le 72$, then we may choose $G = H_1$. We may therefore assume that $n \ge 72$.

Let L be a long lines through q. Then L meets at least n^2+n+1 lines and therefore L misses at most $2n-3$ lines. In view of $k_H \ge 1+n/2$ and because each point p of H lies on r_p-n-1 lines which miss L, it follows that H has at least two points $\ne q$ which have degree less than n+5. Since there is at most one line of degree n (Lemma 9.5), we see that H has a point $p \ne q$ with $r_p = n+1+x$ for some $x \le 3$ and such that p does not lie on a line of degree n.

By part b), the point p lies on at least $f := d-(s-1)x \ge n/6 - 9$ short lines. If $n+1-e$ is the maximal degree of these short lines, then $v-1 \le (r_p-f)n+f(n-e)$, which shows that $fe \le xn$. It follows that $e(n/6-9) \le xn \le 3n$. In view of $n \ge 72$, it follows that $e \le 72$. Therefore p lies on a short line of degree at least n-72. Since p does not lie on n-lines, this completes the proof of part c).

d) Denote the number of short lines through q' by s'+1. We may assume that $s \ge s'$. Since q has degree n+2, it follows from Part b) that q' lies on at least $d+1-s$ short lines, i.e. $s' \ge d-s$. This implies that $2s \ge s+s' \ge d$. Since $(s+1)d \le n$ by part a), we conclude that $2n \ge 2d(s+1) > 2ds \ge d^2$. Hence $d < \sqrt{2n}$ and the lemma is proved.■

<u>9.8 Lemma</u>. It exists a point of degree n+2 which does not lie on an n-line.

Proof. Let p be a real point. Since there exists at most one n-line, it suffices to show that some line through p has two points of degree n+2. Assume to the contrary that every line through p has at most one point of degree n+2. For every line X through p, let t_X be the number of non-real points on X. Then the excess of a line X through p is at least $2t_X-1$. We chose a line L through p for which t_L is maximal. If X is a any line through p, then Lemma 9.3 shows that $b \geq n^2+n+1+2t_L-1+2t_X-1$. Since $b < n^2+3n-1$, we obtain $t_L+t_X \leq n$. The maximility of t_L implies now that every line through p has at most n/2 non-real points. Hence there are at most $(n+1)n/2$ non-real points. This contradicts Lemma 9.6.■

9.9 Lemma. If $b < n^2+3n-143$ then $b \geq n^2+3n+1-2\sqrt{(2n)}$ and there exists a line H satisfying $n+1-\sqrt{(2n)} \leq k_H \leq n-1$.

Proof. In view of $b < n^2+3n-143$, Lemma 9.4 b) shows that there does not exist a line L satisfying $n-71 \leq k_L \leq n-1$.

Consider the set M of points of degree n+2 which do not lie on a line of degree n, and the set M' of short lines containing a point of M. The preceding lemma shows that M and hence M' is not empty.

Let H be a line of maximal degree of M'. Part c) of the Lemma 9.7 shows that H has three points of degree n+2. Since there is at most one line of degree n (Lemma 9.5), it follows that H has two points of degree n+2 which do not lie on a line of degree n. Lemma 9.7 d) shows now that H has degree at least $n+1-\sqrt{(2n)}$. Lemma 9.4 b) shows that there are at least $n^2+3n+1-\sqrt{(2n)}$ lines.■

Now also the proof of Theorem 9.1 is completed. We have even shown something more. We know that there are short lines which have degree at least $n+1-\sqrt{n}$. Probably this information can be used to show that we have actually $b \geq n^2+3n$. There are two questions which seems to be easier. If $b \leq n^2+3n$, is it true that there is no n-line, and is it true that every long line meets every other long line?

An interesting property for linear spaces with $v = n^2+n+1$ was noticed by Erdős, Fowler, Sós and Wilson (1985). If $n = q^2+q$ for a positive integer $q \gg 1$, then every linear space with v points satisfies $b = n^2+n+1$ (and is a generalized projective plane), or $b = n^2+2n+1$ (and is a complete projectively inflated affine plane) or $b \geq n^2+2n+1+q$. This can be seen as follows. If $n^2+2n+1 < b \leq n^2+2n+1+q$, then a linear space with v points and b lines is an affine plane with a linear space D at infinity (by Theorem 9.1). The space D has $v' := n+1 = q^2+q+1$ points and $b' := b-n^2-n-1$ lines. Since $v' < b'$, the Theorem of Totten shows that $b' \geq q^2+2q+1$ so that $b \geq n^2+n+1+b' \geq n^2+2n+1+q$.

10. A hypothetical structure

In this section we shall study a special class of linear spaces which is related to many problems in the next sections and also to the Theorem of Totten.

10.1 Definition. Suppose that n and d are integers satisfying $1 \leq d \leq n-2$. Then an $L(n,d)$ is an $(n+1,1)$-design with n^2-d points which has only lines of degree $n-d$ and n. The lines of degree n are called *long* and the lines of degree $n-d$ are called *short*.

We shall first collect basic properties of these linear spaces and describe the known examples. In particular, we shall show that for each $L(n,d)$, there is an integer $z \geq 0$ with $z(n-d) = d(d-1)$. We shall see that an $L(n,d)$ is a punctured affine plane of order n if $z = 0$ and the complement of a Baer-subplane in a projective plane of order n if $z = 1$. For $z \geq 2$, no examples of an $L(n,d)$ are known and we give a non-existence criterion in this case.

Then we show that for any positive integers n, d, and z with $z(n-d) = d(d-1)$ and $z \geq 2$, every linear space with maximal point degree $n+1$ and n^2+n+z lines has at most n^2-d points with equality if and only if it is an $L(n,d)$.

We shall use the linear spaces $L(n,d)$ to construct new ones. The new spaces will be related to problems in the next sections. For example, in section 12 we shall complete the classification of linear spaces which have a point of degree at most n and satisfy $n^2-n+2 \leq v \leq b \leq n^2+n+1$. The case $v = n^2-n+2$ and $b = n^2+n+1$ was left unsettled in section 4. We shall see that every example of such a linear space with a point of degree n is related to an $L(n,d)$ with $z = 2$.

Our first lemma collects basic properties of an $L(n,d)$.

10.2 Lemma. Suppose L is an $L(n,d)$.

a) Every point of L lies on a unique short and on n long lines.

d) The short lines form a parallel-class Π.

c) The number z defined by $z(n-d) = d^2-d$ is an integer.

d) L has n^2-d lines of degree n and $n+d+z$ lines of degree $n-d$.

e) Every long line N lies in a unique parallel-class π_N, which consists of n−d long lines and d+z short lines.

f) If $z = 0$, then L is a punctured affine plane of order n, and if $z = 1$, then L is the complement of a Baer-subplane in a projective plane of order n.

Proof. Part a) follows from $v = n^2-d$, and b) follows from a). Part b) implies that $|\pi|(n-d) = v = n^2-d$. It follows that $|\pi| = n+d+z$ where z is defined as in part c). Since each point lies on n long lines, the number of long lines equals to $v = n^2-d$. This proves d).

Let N be a long line. Since every point outside of N lies on a unique line parallel to N, N is contained in a unique parallel-class, which consists of N and all the lines parallel to N. Since N meets n short lines and n(n−1) long lines, it follows from d) that π_N consists of d+z short and n−d long lines.

It remains to prove part f). If $z = 0$, then $d = 1$ and L is obviously a punctured affine plane of order n. If $z = 1$, then $n = d^2$ and L is the pseudo-complement of a Baer-subplane in a projective plane of order n. Theorem 5.8 shows that L is in fact a complement in this case.■

For $z \geq 2$, no example for an L(n,d) with $z(n-d) = d(d-1)$ is known. However we can give a non-existence criterion. Therefore we need the following definition.

10.3 Definition. a) Suppose v,k,μ are integers with $2 \leq k \leq v-2$. A *2-(v,k,μ) block design* is an incidence structure with v points in which every line has degree k and any two distinct points are contained in exactly μ lines.

b) A 2-(v,k,μ) block design is called *symmetric*, if every point has degree k.

Examples. An affine plane of order n is a $2-(n^2,n,1)$ block design, and the projective planes are the symmetric block designs with $\mu = 1$.

10.4 Lemma. If there is an L(n,d) with $z(n-d) = d(d-1)$ and $z \geq 1$, then it exists a symmetric 2-(n+d+z,d+z,z) block design.

Proof. Suppose L is an L(n,d) with $z(n-d) = d(d-1)$. By Lemma 10.2, each n-line N is contained in a unique parallel class π_N, which consists of n−d long and d+z short lines. We call π_N a *clique*. It consists of N and the lines which miss N.

Let Π_N and Π_G be two cliques. If $G \in \Pi_N$, then every line of Π_N is parallel to G so that $\Pi_N = \Pi_G$. If $G \notin \Pi_N$, then G meets N and k_G-1 other lines of Π_N, so $|\Pi_N \cap \Pi_G| = z$. In this case, each lines L of $\Pi_N \cap \Pi_G$ has degree n−d, since the point $N \cap G$ is contained in two lines parallel to L. Hence, any two distinct cliques have z lines in common, which are short. Since there are $n^2-d = (n+d+z)(n-d)$ long lines, and because each clique has n−d long lines (Lemma 10.2 e)), it follows that L has n+d+z cliques.

A short line L misses $n^2-d-k_Ln = n^2-d-(n-d)n = (d+z)(n-d)$ long lines, so it is contained in d+z cliques. If H is a second short line, then H misses L and meets k_Hd long lines which miss L. Hence, there are $(d+z)(n-d)-k_Hd = z(n-d)$ long lines which miss H and L. It follows that there are z cliques which contain H and L.

We have shown that any two of the n+d+z lines of degree n−d are contained in z cliques and that every clique has d+z lines of degree n−d. Thus, the cliques induce a 2−(n+d+z,d+z,z) block design on the set of lines of degree n−d. Since every short line is contained in d+z cliques, this design is symmetric.■

Two non−existence criteria for symmetric 2−(v,k,μ) block designs are known. The first one is easy to prove.

<u>10.5 Lemma</u> (Schützenberger, 1949). Suppose it exists a symmetric 2−(v,k,μ) block design and v is even. Then k−μ is a perfect square.

<u>Proof</u>. Let D be a symmetric 2−(v,k,μ) block design. Then every point and line has degree k, so the basic equation for incidence structures implies that there are v lines. Let A be any incidence matrix of D. Then A is a square matrix, since D has the same number of points and lines. Since every point has degree k and because any two points are contained in μ lines, the matrix $C := (c_{ij}) := AA^t$ satisfies $c_{ii} = k$ and $c_{ij} = \mu$, $i \neq j$. It follows easily that $\det(C) = k^2(k-\mu)^{v-1}$. In view of $\det(C) = \det(AA^t) = \det(A)^2$ and because v is even, this implies that k−μ is a perfect square.■

If v is not even, the situation is much more difficult. A famous non−existence criterion has been given by Chowla and Ryser (1950). A selfcontained proof may be found in chapter 4 of (Beutelspacher, 1982).

10.6 Result (Chowla and Ryser, 1950). Suppose there exists a symmetric 2-(v,k,μ) with an odd number v of points, and suppose that p is an odd prime which divides the square-free part of $k-\mu$ and which does not divide μ.

a) If $v \equiv 1 \bmod 4$, then there is an integer x with $x^2 \equiv \mu \bmod p$.

b) If $v \equiv 3 \bmod 4$, then there is an integer x with $x^2 \equiv -\mu \bmod p$.■

A non-existence criterion for an $L(n,d)$ with $z(n-d) = d(d-1)$ and $z \geq 1$ can now be obtained from Result 10.6, Lemma 10.5 and Lemma 10.4.

The case $\mu = 2$ will be particularly interesting for us. Therefore we translate Result 10.6 in this case in another form. But first we need the following lemma.

10.7 Lemma. Suppose p is an odd prime.

a) If there exists an integer x with $x^2 \equiv 2 \bmod p$, then $p \equiv 1$ or $p \equiv 7 \bmod 8$.

b) If there exists an integer x with $x^2 \equiv -2 \bmod p$, then $p \equiv 1$ or $p \equiv 3 \bmod 8$.

Proof. This follows from the Lemma of Gauß, which can be found in almost every book about number theory.■

10.8 Corollary. Suppose there exists a symmetric 2-$(v,d+2,2)$ block design.

a) If $d \equiv 0$ or $d \equiv 1 \bmod 4$, then d is a perfect square.

b) If $d \equiv 2$ or $d \equiv 3 \bmod 8$, then no prime with $p \equiv 5$ or $p \equiv 7 \bmod 8$ divides the square free part of d.

c) If $d \equiv 6$ or $d \equiv 7 \bmod 8$, then no prime with $p \equiv 3$ or $p \equiv 5 \bmod 8$ divides the square free part of d.

Proof. Each point p has degree $k := d+2$, since the design is symmetric. Since p is joined to every point by two lines, we have $2(v-1) = k(k-1) = (d+1)(d+2)$.

a) If $d \equiv 0$ or $d \equiv 1 \bmod 4$, then 2 divides v. It follows therefore from 10.5 that d is a perfect square.

b,c) If $d \equiv 2$ or $d \equiv 3 \bmod 8$, then $v \equiv 3 \bmod 4$. If $d \equiv 6$ or $d \equiv 7 \bmod 8$, then $v \equiv 1 \bmod 4$. b) and c) follow therefore from 10.6 and 10.7.■

For $d \leq 25$, only 3, 4, 7, 9, 11 , 14, 16, 18, 19, and 25 satisfy the conditions a), b), and c) of 10.8. The corresponding values for n are 6, 10, 28, 45, 66, 105, 136, 171, 190, 276, and 325. It is known that there is a symmetric 2-$(v,d+2,2)$ design for $d \in \{3,4,7,9,11\}$ (see Beth, Jungnickel, Lenz, 1985). However the existence of a symmetric 2-(v,d+2,2) block design does not imply that an L(n,d) for the corresponding n exists. It seems even not very likely that there are any. In section 13, we shall show that $n = 6$ is not possible although there is a symmetric 2-(11,5,2) block design.

In the following theorem we give a characterization of the spaces L(n,d). Notice that this theorem improves a part of the Theorem of Totten.

10.9 Theorem. Suppose that L is a linear space with $b \geq n^2+n+2$ lines in which every point is on at most n+1 lines. Denote by z the unique integer with $b = n^2+n+z$ and by e the unique real positive number with $z(n-e) = e(e-1)$. Then $v \leq n^2-e$ with equality if and only if L is an L(n,e).

Proof. We set

$$v = n^2-f \qquad \text{and} \qquad T = \sum_{p \in \mathcal{P}} (n+1-r_p).$$

In order to prove the theorem, we shall assume that $f \leq e$ and we have to show that equality holds.

First we want to show that every line has degree at most n. Assume to the contrary that there is a line L of degree n+1. Since n+1 is the maximal point degree, the line L meets at most n^2+n other lines and there does not exist a line which misses L. But the number of lines is n^2+n+2, a contradiction.

Consequently, every line has degree at most n. Hence, if L is any line and p any point of L, then

$$(+) \qquad n^2-f = v \leq k_L+(r_p-1)(n-1) \leq k_L+n(n-1)$$

so that $k_L \geq n-f$. Now set $e_L = n-k_L$ for every line L. Then we have shown that $0 \leq e_L \leq f$ and therefore $e_L(f-e_L) \geq 0$ for every line L. We shall show that we have in fact $e_L = 0$ or $e_L = f$.

Since every line has degree at most n, we have $n \geq 2$. Therefore $z(n-e) = e(e-1)$ implies that $e > 1$. It follows that

$$z(2n-1-f) \geq z(2n-1-e) = 2z(n-e)+z(e-1) = 2e(e-1)+z(e-1) > 0.$$

Therefore $2n-1-f > 0$.

Using $e_L(f-e_L) \geq 0$ for every line L and the following two equations

$$v(v-1) = \sum_{L\in \mathcal{L}} k_L(k_L-1) \quad \text{and} \quad v(n+1)+T = \sum_{p\in \mathcal{P}} r_p = \sum_{L\in \mathcal{L}} k_L,$$

we obtain

$$\begin{aligned} 0 \leq \sum_{L\in \mathcal{L}} e_L(f-e_L) &= \sum_{L\in \mathcal{L}} (n-k_L)(f+k_L-n) \\ &= \sum_{L\in \mathcal{L}} [-k_L(k_L-1)+ (2n-1-f)k_L - n(n-f)] \\ &= -v(v-1) + (2n-1-f)[v(n+1)+T] - bn(n-f) \\ &= -(n^2-f)(n^2-1-f) + (2n-1-f)[(n^2-f)(n+1)+T] - (n^2+n+z)n(n-f) \\ &= (2n-1-f)T + n[f(f+z-1)-zn] \\ &= (2n-1-f)T + n[f(f+z-1)-e(e+z-1)]. \\ &= (2n-1-f)T + n(f-e)(z-1) + n(e^2-f^2). \end{aligned}$$

Since every point has degree at most n+1, we have $T \leq 0$. Since we have also $z-1 \geq 1$ and $0 \leq f \leq e$ ($0 \leq f$ follows from (+)), we obtain $T = 0$, $e = f$, and $e_L(f-e_L) = 0$ for every line L. The equations $T = 0$ means that every point has degree n+1, and $0 = e_L(f-e_L) = e_L(e-e_L)$ means that every line has degree n or n−e. Now $v = n^2-f = n^2-e$ implies that every point lies on a unique line of degree n−e and on n lines of degree n. By definition, L is an L(n,e).■

Now we want to use an L(n,d) to construct new linear spaces. There are several possibilities to do this.

<u>10.10 Definition</u>. Suppose L is an L(n,d) and denote by Π any parallel-class of L. Then the linear space $L\infty\Pi$ is called a *extended* L(n,d). It is obtained from L, if we let the lines of Π intersect in a new point. If Π is the parallel-class consisting of the short lines of L, then $L\infty\Pi$ is called a *closed* L(n,d).

It follows from Lemma 10.2 that the closed L(n,d) with $z = 0$ are the affine planes, and that the closed L(n,d) with $z = 1$ are the closed complements of a Baer-subplane in a projective plane.

If $z = 2$, then an extended L(n,d) has n^2+n+2 lines and n^2-d+1 points where $2n = d(d+1)$. In the next section we shall prove the following. If $n \geq 6$ is an integer and if d is the positive number with $2n = d(d+1)$, then every linear space with n^2+n+2 lines has at most n^2-d+1 points with equality if and only if it is an extended L(n,d). Theorem 10.9 will be an important aid in the proof.

Now suppose that L is an L(n,d) and that Π is a parallel-class containing a long line. Suppose furthermore that $2n = d(d+1)$, i.e. $z = 2$, and consider the extended space $L\infty\Pi$. Then every short line of L which is not contained in Π has still degree n−d in $L\infty\Pi$. Let L be such a short line and let q be a point of L. If we remove L and all its points other than q from $L\infty\Pi$, then we obtain a space L' with n^2-n+2 points and n^2+n+1 lines. Furthermore the point q has degree n in L'. We shall show in section 12 that L' is characterized by these properties (see Theorem 12.1). A first step towards this characterization is made in the following theorem.

<u>10.11 Theorem</u>. Let L be an incidence structure such that any two distinct points are contained in a unique line and every line has degree at least 1. Suppose that there is an integer $n \geq 3$ such that L has a point q of degree n which is contained in n lines of degree n and suppose that every other point has degree n+1. If L has $v = n^2-n+1$ points and $b = n^2+n+1$ lines, then $2n = d(d+1)$ for an integer d and L can be obtained from an L(n,d) by removing a line of degree n−d with all but one of its points.

<u>Proof</u>. Call the lines of degree at most n−1 *short*. For every n-line N with $q \notin N$, we set $M_N = \{N\} \cup \{L \mid L$ is a line parallel to $N\}$ and call M_N a *clique*. We shall see soon that there are cliques. By s+1 we denote then the maximal number of short lines contained in a clique. We prove Theorem 10.12 in several steps.

Step 1. Every line has degree at most n *and every point of degree* n+1 *is contained in at least two* n*-lines.*
For: By the hypotheses, every line through q has degree n. Because q has degree n, every line not containing q has degree at most n. Since there are $n^2-n+1 = (n+1)(n-2)+3$ points, every point of degree n+1 is therefore contained in at least two n-lines.

Step 2. a) Each clique M consists of $n+1$ *mutually parallel lines and* q *is the only point which lies not on any line of M. In particular,* $\sum_{X\in M} k_X = v-1 = n^2-n$.
b) There exists at least two cliques.
c) Any two distinct cliques have a unique line in common and this line is short.
For: a) Let $M = M_N$ be a clique. Since every point of N has degree $n+1$, N is parallel to n lines, which proves that $|M| = n+1$. Since q has degree n, it is not contained in a line parallel to N. Every other point has degree $n+1$ and is therefore contained in a unique line of M.

b) By Step 1, there is an n-line N with $q \notin N$. N determines a clique M. By part a), not every line of M has degree n. Let p be any point which is contained in a short line L of M. Again by part a), p is also contained in an n-line N' which determines a clique M'. In view of $L \notin M'$, we have $M \neq M'$.

c) Suppose M_N and M_G are cliques. If $N = G$ or if N and G are parallel, then $M_N = M_G$. If N and G intersect in a point p, then G intersects n of the lines of M_N so that $M_N \cap M_G$ consists of a single line S. Since p is contained in two lines parallel to S, S is a short line.

Step 3. $s \geq 0$ *and every line has degree at least* $n-s$.
For: By Step 2, not every line of a clique has degree n. This implies that $s \geq 0$. Hence, every n-line has degree at least $n-s$.

Now let X be any short line and denote by p a point of X. By Step 1, p is contained in an n-line N with $q \notin N$. N determines the clique M_N. If G is an n-line of $M_N - \{N\}$, then N is the unique line through p which is parallel to G, so G meets X. Hence X meets every n-line of M_N. Since (by definition) $n-s$ is the minimal number of n-lines contained in a clique, we obtain $k_X \geq n-s$.

Step 4. Let M be a clique with $s+1$ *short lines, let* L *be a short line of maximal degree of M, and* set $k_L = n-d$. *Then* $d(s+1) \leq 2n$ *with equality if and only if every short line of M has degree* $n-d$.
For: This follows from $\sum_{X\in M} k_X = n^2-n$ and $|M| = n+1$ (see Step 2).

Step 5. $2n = s(s+1)$ *and there are at most* $s+2$ *cliques. Every clique has exactly* $s+1$ *short lines. Every short line of a clique has degree* $n-s$ *and is contained in at most one other clique.*

For: Let $\mathcal{M}$ be a clique with the maximal number s+1 of short lines, let L be a short line of maximal degree of $\mathcal{M}$, and set $k_L = n-d$. Furthermore, denote by $N_1,\dots,N_d$ the lines through q which miss L, set $\pi_j = \{N_j\}\cup\{X \mid X \text{ is parallel to } N_j\}$, and $\mathcal{T} := \pi_1\cup\dots\cup\pi_d$.

Since every point outside of N_j has degree n+1, π_j is a parallel-class. We have $L \in \pi_j$ and $|\pi_j| = n+2$. If $j,k \in \{1,\dots,d\}$ with $j \neq k$, then N_k intersects n lines of π_j, so $\pi_j\cap\pi_k$ contains two lines, one of which is L. It follows that $|\mathcal{T}| \geq d(n+1)-\frac{1}{2}d(d-1)+1$ with equality if and only if L is the only line which lies in more than two of the parallel-classes π_j. Since π_j is a parallel class, every line of $\mathcal{T}-\{L\}$ is parallel to L.

Let N be an n-line of $\mathcal{M}$. Then $\mathcal{M} = \mathcal{M}_N$ and N meets every line passing through q. Hence N meets n lines of π_j, so $\mathcal{M}$ and π_j have two lines is common. One of these two lines is L. It follows that $\mathcal{M}$ contains at least n-d lines which are not in $\mathcal{T}$. This shows that $|\mathcal{T}\cup\mathcal{M}| \geq |\mathcal{T}|+n-d \geq (d+1)n-\frac{1}{2}d(d-1)+1$. Obviously, every line of $\mathcal{T}\cup\mathcal{M}-\{L\}$ is parallel to L.

Let H be the line $\neq$ L of $\pi_1\cap\mathcal{M}$. If p is a point of H, then p lies on d+1 lines which miss L. Each of the sets $\mathcal{M},\pi_1,\dots,\pi_d$ contains one of these d+1 lines. Since $H \in \pi_1\cap\mathcal{M}$, it follows that p lies on a line which misses L which is not contained in $\mathcal{T}\cup\mathcal{M}$. Thus, there are k_H lines which miss L and which are not in $\mathcal{T}\cup\mathcal{M}$. Since L misses (d+1)n lines, we obtain $(d+1)n \geq |\mathcal{T}\cup\mathcal{M}-\{L\}|+k_H \geq (d+1)n-\frac{1}{2}d(d-1)+k_H$, i.e. $2k_H \leq d(d-1)$. Together with Step 3 and 4 this implies

$$2(n-s) \leq 2k_H \leq d(d-1) = d(s+1) + d(d-2-s) \leq 2n + d(d-2-s).$$

We obtain $(s-d)(d-2) \leq 0$. In view of $k_H \geq 1$ and $2k_H \leq d(d-1)$, we have $d \geq 2$, and in view of $n-d = k_L \geq n-s$ (Step 3), we have $s \geq d$. Hence, equality holds in the above inequalities. Thus, $2n = d(s+1)$ and $2(n-s) = 2k_H = d(d-1)$. Step 4 and $2n = d(s+1)$ show that every short line of $\mathcal{M}$ has degree n-d. Since H is a short line of $\mathcal{M}$ and in view of $k_H = n-s$, it follows that $s = d$ and $2n = s(s+1)$.

We have also shown that there are exactly $k_H = n-s$ lines which are parallel to L and which are not contained in $\mathcal{T}\cup\mathcal{M}$. Suppose that $\mathcal{M}'$ is another clique containing L. Since s+1 is the maximal number of short lines, $\mathcal{M}'$ has at least n-s long lines. These long lines meet every line passing through q and are therefore not contained in $\mathcal{T}$. Since $\mathcal{M}$ and $\mathcal{M}'$ have only the line L in common (Step 2 c)), it follows that M' has exactly n−s long lines and that every line parallel to L is contained in $T \cup M' \cup M$. The same argument shows now that L is not contained in a third clique.

Since L was any short line of maximal degree of M and because every short line of M has the same degree, we can show in the same way that every short line of M is contained in at most one other clique and that the second clique has s+1 short lines in this case. Because every clique $\neq M$ has a short line in common with M (Step 2), it follows that every clique has s+1 short lines, and that there are at most s+2 cliques with equality if and only if every short line of M is contained in two cliques. As for M, we can show now that every short line of a clique has degree n−s.

We remark that $2n = s(s+1)$ implies that $s > 1$, since every line has degree at least one and because there exists lines of degree n−s.

Step 6. The number of cliques is s+2. *If* S *is a short line which lies in two cliques, then every point of* L *lies on one line of degree* n−s *(which is* L*), on* s+1 *lines of degree* n, *and on* n−1−s *lines of degree* n−1.

For: By Step 2 there are two cliques, which have a short line in common. Let S be any short line which is contained in two cliques and let p be a point of S. Then p lies on the n-line pq. Every other n-line through p determines a clique which certainly does not contain S. Since S is contained in two cliques and because the number of cliques is at most s+2, it follows that p lies on at most s+1 lines of degree n. Since p lies on the line S of degree n−s (Step 5) and in view of $v = n^2-n+1$, it follows that p lies on exactly in s+1 lines of degree n and on n−1−s lines of degree n−1 (count the number of points from p and use $2n = s(s+1)$). In particular, the number of cliques is s+2.

Step 7. Every point $\neq$ q *lies on one line of degree* n−s, *on* s+1 *lines of degree* n, *and on* n−1−s *lines of degree* n−1.

For: By Step 5, every clique has s+1 short lines, and every short line lies in at most two cliques. Since distinct cliques have a short line in common (Step 2) and because the number of cliques is s+2 (Step 6), it follows that every short line is either contained in no or in two cliques. Since each of the s+2 cliques has s+1 short lines, we conclude that there are exactly $\frac{1}{2}(s+1)(s+2)$ short lines which lie in two cliques. Step 6 shows that these lines are mutually parallel. In view of $\frac{1}{2}(s+1)(s+2)(n-s) = n^2-n = v-1$, it follows that every point $\neq q$ lies on a short line which is contained in two cliques. Now Step 7 follows from Step 6.

Step 8. There are $n(n-1-s)$ *lines of degree* $n-1$, *and every line of degree* $n-1$ *is contained in a unique set consisting of* n *mutually parallel lines of degree* $n-1$.
For: It follows from Step 7 that the number of lines of degree $n-1$ is $(n-1-s)n$. Let H be a line of degree $n-1$, let N the unique line through q missing H, and denote by $\pi_N = \{N\} \cup \{X \mid X \text{ is parallel to } N\}$ the parallel-class containing N (and H). Since every point p with $p \notin \{q\} \cup H$ lies on two lines parallel to H, the set π consisting of H and the lines parallel to H which are not in π_N is a set of mutually parallel lines. Since H misses 2n lines and in view of $|\pi_N| = n+2$, we have $|\pi| = n$. Furthermore, the lines of π cover the n^2-n points $\neq q$. Since π does not contain an n-line (Step 5), it follows that every line of π has degree $n-1$.

Now let π' be any set which contains H and which consists of n mutually parallel lines of degree $n-1$. Since q lies only on n-lines, no line of π' contains q. Consequently, the lines of π' cover the n^2-n points $\neq q$. In particular, each point of $N-\{q\}$ lies on a line of π'. Since each of point $N-\{q\}$ lies on a unique line other than N which is parallel to H, the set π' is uniquely determined. It consists of the line H and the $n-1$ lines $\neq N$ which miss H and meet N.

Step 9. L *can be obtained from an* L(n,s) *by removing a line of degree* $n-s$ *with all but one of its points.*
For: Step 8 shows that the set of the lines of degree $n-1$ can be partitioned uniquely into sets $\pi_1,\ldots,\pi_{n-s-1}$ of n mutually parallel lines such that each point other than q lies on a unique line of each set π_j.

We extend L in the following way. We adjoin an infinite point p_j to the lines of π_j for all j and then we adjoin the line $\{q,p_1,\ldots,p_{n-1-s}\}$. In this way, we obtain an (n+1,1)-design L' with n^2-s points and n^2+n+2 lines each of which has degree $n-s$ or n. Since the new line has degree $n-s$, the lines of degree $n-s$ form a parallel-class of L' (Step 7). By definition, L' is an L(n,s). By our construction, L is obtained from L' by removing the new line $\{q, p_1, \ldots, p_{n-1-s}\}$ with all its points other than q. This completes the proof of Step 8 and the theorem.■

We have already pointed out that the spaces L(n,d) are related to some problems in the next sections. This is the reason why we try to find non-existence criteria for them. However an L(n,d) is also interesting from other points of view. We mention one of them.

Suppose I is an incidence structure with n^2 points such that any two distinct points are contained in at least one line and every line has degree n. Then it is obvious that $b \geq n^2+n$ with equality if and only if I is an affine plane of order n. Now suppose that $b = n^2+n+1$. Such a structure has been called a *failed affine plane* (Mendelsohn and Assaf, 1987). An example for I can be obtained from an affine plane A of order n by adjoining a set of n points, considered as a new line, to A. Such an example is called *trivial* failed affine plane. A non-trivial example can be obtained from the complement of a Baer-subplane in a projective plane of order $n = m^2$ by adjoining m new points to each Baer-line. Baker, Blokhuis, Brouwer and Wilbrink (1990) showed that every other non-trivial example can be constructed from some L(n,d) with $2n = d(d+1)$ (here $n = d+1 = 2$ is allowed) in the following way. First let the lines of each parallel class π_N, N an n-line, intersect in an infinite point to obtain a structure D with n^2+n+2 points and lines. (Notice that the structure induced on the infinite points by the lines of previous length n−d form now a Biplane which is dual to the one constructed in the proof of Lemma 10.4). Now remove a line of degree n+2 from D. This results in a failed affine plane.

11. Linear spaces with n^2+n+2 lines.

In the first sections we have studied linear spaces with at most n^2+n+1 lines and $v \geq n^2-n+2$ points and we proved embedding results for such spaces. Then we proved the Theorem of Totten, a part of which characterizes linear spaces satisfying $v \leq n^2+n+1 < b$ and $b \leq v+n$. It turned out that these are inflated affine planes or Lin's Cross. Notice that an inflated affine plane always has at least n^2+n+3 lines, since the affine plane has n^2+n lines and the linear space at infinity has at least three lines. Hence, a non-degenerate linear space with n^2+n+2 lines is either Lin's Cross or has $v \leq n^2+1$ points. Stinson (1983) showed that $v = n^2+1$ can not occur for $n \geq 4$. It is natural to asked for the maximum possible number of points a linear space with n^2+n+2 lines can have. A lower bound is n^2-n+4 as is shown by the following example.

Let L' be an affine plane of order n with a triangle at infinity, so L' has n^2+3 points and n^2+n+3 lines. Remove an n-line and all its points except one from L' ($n \geq 3$). This gives a linear space L with n^2-n+4 points and n^2+n+2 lines. For $n = 3$, we have $v = n^2+1 = 10$ and $b = n^2+n+2 = 14$ and L is the unique linear space with 10 points and 14 lines (Stinson, 1983).

Recall that we studied a class of linear spaces, which we called L(n,d) or extended L(n,d) (see Definition 10.1 and 10.10). We saw in Lemma 10.2 that they have n^2+n+z lines and $v = n^2-d$ points where z is the integer defined by $z(n-d) = d(d-1)$. However, for $z \geq 2$, we do not know examples for such linear spaces. Eventhough they are useful in the following theorem.

<u>11.1 Theorem</u> (Metsch, 1991b). Suppose L is a non-degenerate linear space with n^2+n+2 lines for some integer $n \geq 6$. Denote by v its number of points and by e the positive number with $2n = e(e+1)$. Then we have:

a) $v \leq n^2+1-e$ with equality if and only if L is an extended or closed L(n,e).

b) If every point has degree at most n+1 then $v \leq n^2-e$ with equality if and only if L is an L(n,e).

Roughly speaking, this theorem says that every non-degenerate linear space with n^2+n+2 lines, $n \geq 6$, has at most $n^2+1-\sqrt{(2n)}$ points. The above examples show that Theorem 11.1 does not remain true for $n \leq 5$.

Part b) of Theorem 11.1 has already been proved in Theorem 10.9. For the proof of part a) we use the following notation.

Throughout in this section we denote by $\mathbf{L} = (p, L)$ a non-degenerate linear space with $b = n^2+n+2$ lines for some integer $n \geq 6$, v denotes its number of points, and e denotes the unique positive number with $2n = e(e+1)$. Furthermore, we define

$$T = \sum_{p \in p} (r_p - n - 1), \qquad f = n^2 - v, \qquad \text{and}$$

$$t_L = \sum_{p \in L} (r_p - n - 1) \quad \text{and} \quad d_L = n - k_L \quad \text{for every line } L.$$

Part a) of Theorem 11.1 will be proved in several lemmas. We shall assume throughout that $\mathbf{L}$ has at least n^2+1-e points, which implies that $f \leq e-1$.

<u>11.2 Lemma</u>. $\sum_{L \in L} d_L(f-d_L) = (2n-f-1)T - n[2n-f(f+1)]$.

<u>Proof</u>. The equations (B_1) and (B_2) in Section 1 show that $\$k_L = \$r_p = v(n+1)+T$ and $\$\, k_L(k_L-1) = v(v-1) = (n^2-f)(n^2-f-1)$. Using $k_L = n-d_L$ for every line L, the equation in Lemma 11.2 is easily proved.■

<u>11.3 Lemma</u>. Every line has degree at most $n+1$.

<u>Proof</u>. Assume to the contrary that there is a line L with $k := k_L \geq n+2$. Since $\mathbf{L}$ is not a near-pencil, $k \leq v-2$. The Lemma of Stanton and Kalbfleisch (Lemma 2.4) implies therefore that

$$b \geq f(k,v) \geq \max\{f(n+2,v), f(v-2,v)\}.$$

We recall that $f(k,v) = 1+k^2(v-k)/(v-1)$ for $2 \leq k \leq v-1$.
First consider the case $b \geq f(n+2,v)$. Then $(n^2+n+1)(v-1) \geq (n+2)^2(v-n-2)$ so that $0 \geq (n+1)[3(v-1)-(n+2)^2]$. It follows that $4+3e \geq 2n(n-2) = (e+1)(e+2)(n-2)$, since $v \geq n^2-e+1$. But this contradictis $n \geq 6$ and $2n = e(e+1)$.
Now suppose that $b \geq f(v-2,v)$. Then $(n^2+n+1)(v-1) \geq 2(v-2)^2$, so that $n^2+n+1 \geq (v-2)[2(v-2)-n^2-n-1] \geq (v-2)[(v-2)-n-e-2]$, In view of $n^2+n+1 \leq v+n+e$, this implies that $n+e+2 \geq (v-2)[(v-2)-n-e-3]$. It follows that $v-2 \leq n+e+4$, which gives a contradiction as before.■

<u>11.4 Lemma</u>. a) If an n-line N is parallel to two intersecting lines H and L then $(d_H+1)(d_L+1) \geq n-t_N$ and $d_H+d_L+2 \geq 2\sqrt{(n-t_N)}$.

b) Every n-line meets every line of degree n+1.

c) If $\mathcal{M}$ is the set of lines parallel to an n-line N, then $|\mathcal{M}| = n+1-t_N$ and $\sum_{L\in\mathcal{M}} k_L = n^2-n-f+(T-t_N)$.

<u>Proof</u>. Let N be an n-line and denote by $\mathcal{M}$ the set of lines which miss N. If two lines H and L of $\mathcal{M}$ intersect, then the Parallel-Lemma (see Section 1) shows that $(d_H+1)(d_L+1) \geq n-t_N$, which implies that $d_H+d_L+2 \geq 2\sqrt{(n-t_N)}$.

Since $b = n^2+n+2$, we have $|\mathcal{M}| = n+1-t_N$. Counting the number of incident point-line pairs (p,L) with $L \in \mathcal{M}$ in two different ways gives the equation in part c), since every point p outside of N lies in r_p-n lines of $\mathcal{M}$ and because $\sum_{p\notin N} (r_p-n) = v-k_N+(T-t_N)$ (see equation (B_4) in chapter 1).

Assume by way of contradiction that N is parallel to a line L of degree n+1. Then every point of N has degree at least n+2, which implies that $t_N \geq n$. In view of $v > 2n+1 = k_L+k_N$, there exists a point p outside of L and N. Since p is outside of L, it has degree at least n+1 and lies therefore on a line H parallel to N. We conclude that $|\mathcal{M}| \geq |\{L,H\}| = 2$, which contradicts $|\mathcal{M}| = n+1-t_N \leq 1$. Thus N meets every line of degree n+1.∎

<u>11.5 Lemma</u>. If there is a point q of degree at most n then q has degree n, every other point has degree at least n+1, and q lies on at least $n-1-f \geq n-e \geq 3$ lines of degree n+1.

<u>Proof</u>. If α denotes the number of lines of degree n+1 passing through q then

$$n^2-f = v \leq 1+(r_q-\alpha)(n-1)+\alpha n = 1+r_q(n-1)+\alpha.$$

In view of $\alpha \leq r_q$, we obtain $r_q = n$ and $\alpha \geq n-1-f \geq n-e$. If $e \leq 3$ then $n-e \geq 3$, since $n \geq 6$, and if $e \geq 3$ then $n-e \geq 3$, since $2n = e(e+1)$. Thus q lies on at least 3 lines of degree n+1, which implies that every point other than q has degree at least n+1.∎

<u>11.6 Lemma</u>. If there is a point q of degree n then every line L of degree n+1 passes through q and its n points $\neq$ q have degree n+1.

Proof. Obviously, every line not passing through q has degree at most $r_q = n$. Denote by $L_1,\dots,L_\alpha$ the lines of degree $n+1$ through q. Lemma 11.3 shows that $\alpha \geq n-1-f \geq 3$.

Assume by way of contradiction that L_1 has a point q_1 of degree at least $n+2$. Then L_1 meets at least n^2+n other lines. Let H_2 be a line through q_1 which misses L_2, and let q_3 be a point of $H_2-\{q_1\}$. The Transfer-Lemma (Section 1) shows that q_3 lies on a line H_1 which misses L_1 and meets L_2 in a point q_2. Since L_1 meets n^2+n other lines, H_1 is the only line parallel to L_1. It follows that every point outside of H_1 and L_1 has degree $n+1$ and that every point of H_1 has degree $n+2$. In the same way follows that every point outside of L_2 and H_2 has degree $n+1$, that every point of H_2 has degree $n+2$. Hence $H_1 = \{q_2,q_3\}$ and $H_2 = \{q_1,q_3\}$. and every point $\neq q, q_1, q_2, q_3$ has degree $n+1$.

If X is any line through q which does not contain q_1, q_2, or q_3 then H_1 and H_2 are parallel to X. Because q_3 has degree $n+2$, this implies that X has degree at most n. Since q lies on $\alpha \geq n-1-f \geq 3$ lines of degree $n+1$, we conclude that $\alpha = n-1-f = 3$ and that L_3 passes through q and q_3. Also the line L_3 meets n^2+n other lines and is therefore parallel to a unique line H_3. Since the points of H_3 have degree $n+2$ and because q_1 and q_2 are the only points of degree $n+2$ outside of L_3, we have $H_3 = \{q_1,q_2\}$. Now we have found three 2-lines H_1, H_2, H_3 which form a triangle.

If we replace the lines H_1, H_2, and H_3 by the new line $\{q_1,q_2,q_3\}$ and if we add the new line $\{q\}$ of degree one then we obtain an $(n+1,1)$-design D with n^2+n+1 lines and v points. Furthermore, since $v = n^2-f$, every point of D which lies not on one of the three lines of degree $n+1$ is contained in at least $n+1-f$ lines of degree n. In view of $v+(n+1-f) \geq n^2$, Corollary 3.5 b) implies that D can be extended to a projective plane of order n.

We have already proved that $3 = n-1-f$. In view of $2n = e(e+1)$, $n \geq 6$, and $f \leq e-1$, it follows that $n = 6$. Consequently, D can be embedded into a projective plane of order 6. However, Tarry (1900) and also Bruck and Ryser (1949) proved that there is no projective plane of order 6 (we shall show prove this in Section 13). This contradiction completes the proof of the lemma.■

11.7 Lemma. Suppose that N is a line of degree n with $t_N > 0$. Then $t_N = 1$ and every line missing N has degree at most n−1.

Proof. Let α be the number of points outside of N which have degree at least n+2 and set $T' := T - t_N$. Since there is at most one point of degree n (Lemma 11.6), we have $T' \geq \alpha - 1$ with equality if and only if there is a point of degree n outside of N and if every other point outside of N has degree n+1 or n+2. If M denotes the set of lines parallel to N then Lemma 11.4 shows $|M| = n+1-t_N$ and

$$n^2-n-f+(\alpha-1) \leq n^2-f-n+T' \leq \sum_{L\in M} k_L.$$

If every line of M has degree at most n−1 then we obtain $n^2-n-f+(\alpha-1) \leq (n+1-t_N)(n-1)$ so that $(t_N-1)(n-1) \leq f+1-\alpha \leq e-\alpha \leq e$, which shows $t_N = 1$.

It remains to show that N meets every other n-line (see Lemma 11.4 b)). Assume to the contrary that N is parallel to a line N' of degree n. Since $t_N > 0$, there is a point $p \in N$ with $r_p \geq n+2$. The point p lies on a line $H \neq N$ which is parallel to N'. Let s be a point of $H-\{p\}$. Then s lies also on a line H' which misses N and meets N' (Transfer Lemma). If p' is the point of intersection of N' and H' then p' has degree at least n+2, since it lies on two lines missing N. This implies that $\alpha \geq 1$. Lemma 11.4 a) shows that

$$1+d_{H'} = (1+d_{N'})(1+d_{H'}) \geq n-t_N,$$

i.e. $t_N+1 \geq n-d_{H'} = k_{H'} \geq 2$. In view of $H' \in M$, it follows that

$$n^2-n-f \leq \sum_{L\in M} k_L \leq (n-t_N)n+k_{H'} \leq (n-t_N)n+t_N+1.$$

Hence, $(n-1)(t_N-1) \leq f+2 \leq e+1$. In view of $n \geq 6$ and $2n = e(e+1)$, we obtain $t_N < 2$. Hence $t_N = 1$ and $k_{H'} = 2$.

Since every point of N' has degree at least n+1 (because N' misses N) and since $r_{p'} \geq n+2$, we have $t_{N'} \geq 1$. The same argument shows now that $k_H = 2$ and $t_{N'} = 1$. We have shown that every n-line X of M has a point x of degree at least n+2 and that every line $\neq X$ of M through x has degree 2. Thus if β is the number of n-lines in M then $\alpha \geq \beta$. It follows that

$$n^2-n-f+(\alpha-1) \leq n^2-n-f+T' = \sum_{L\in M} k_L \leq k_{H'}+(n-t_N-\beta)(n-1)+\beta n$$
$$= 2+(n-1)(n-1)+\beta.$$

Hence $n \le f+4+\beta-\alpha \le f+4 \le e+3$. In view of $12 \le 2n = e(e+1)$, we obtain $n = e+3 = 6$ so that $\alpha = \beta$ and $T' = \alpha-1$. $T' = \alpha-1$ means that there is a point q of degree n outside of N and that every other point outside of N has degree n+1 or n+2. By Lemma 11.5, q is contained in at least three lines L_1, L_2, and L_3 of degree n+1. It follows that the 2-line H' is parallel to a line of degree n+1 (notice that $q \notin H'$, since every line through q meets N), which implies that every point of H' has degree at least n+2. Lemma 11.6 shows now that H' is parallel to every line of degree n+1. In the same way follows that H is parallel to every line of degree n+1. Since H and H' intersect in the point s, it follows that s has degree at least n+3. But we have shown that every point outside of N has degree n, n+1 or n+2, a contradiction.■

<u>11.8 Lemma</u>. Every point has degree at least n+1.

<u>Proof</u>. Assume to the contrary that there is a point q of degree n. By Lemma 11.5, q is contained in $\alpha \ge n-1-f$ lines of degree n+1. Furthermore, each point $\ne q$ of one of these α lines has degree n+1. Let L be any line of degree n+1 through q. Then L meets n^2+n-1 other lines, so L is parallel to two lines H and G. Every point of G and H has degree n+2 except when G and H intersect in which case the point of intersection has degree n+3. H is therefore parallel to the α lines of degree n+1 through q. Hence $k_H \le r_q-\alpha \le f+1$. Let p be a point of H which is not on G. If β is the number of n-lines through p then

$$n^2-f = v \le k_H+(r_p-1-\beta)(n-2)+\beta(n-1) = k_H+n^2-n-2+\beta \le n^2-n-1+f+\beta$$

so that $\beta \ge n+1-2f > 1$. Consequently, p lies on a line N of degree n with $q \notin N$. Since q is the only point of degree n, we have $t_N \ge 1$. Lemma 11.7 shows therefore that $t_N = 1$. Hence every point $\ne p$ of N has degree n+1, which implies that N is parallel to G. If $\mathcal{M}$ is the set of lines parallel to N then $G \in \mathcal{M}$ and every line of $\mathcal{M}$ has degree at most n-1 (Lemma 11.7). Lemma 11.4 shows therefore that

$$n^2-n-f+(T-t_N) \le |\mathcal{M}-\{G\}|(n-1)+k_G = (n-t_N)(n-1)+k_G.$$

It follows that $k_G \ge n-1-f+(T-t_N) > T-t_N$. But

$$T-t_N = \sum_{x \notin N} (r_x-n-1) \ge |G \cup H-\{q\}| \ge k_G.$$

This contradiction completes the proof of the lemma.■

<u>11.9 Lemma</u>. Suppose it exists an n-line N with $t_N = 1$. Then every line contains a point of degree n+1.

<u>Proof</u>. In view of $t_N = 1$, N has a point unique q of degree n+2 while every other point of N has degree n+1 (Lemma 11.8). Let M be the set of lines parallel to N, set $T' = T-t_N$, and let α denote the number of points of degree at least n+2 outside of N. Then

$$T' = T-t_N = \sum_{p \notin N} (r_p-n-1) \geq \alpha.$$

From Lemma 11.4 we obtain $|M| = n+1-t_N$ and

$$n^2-n-f+\alpha \leq n^2-n-f+T' = \sum_{L \in M} k_L.$$

Assume by way of contradiction that there is a line H which has no point of degree n+1. First we consider the case that H is parallel to N. Then we have $\alpha \geq k_H$. Since each line of M has degree at most n−1 (Lemma 11.7), we obtain

$$n^2-n-f+k_H \leq \sum_{L \in M} k_L \leq |M-\{H\}|(n-1)+k_H = (n-1)^2+k_H,$$

a contradiction.

Now consider the case that H meets N. Then $H \cap N = \{q\}$ and $\alpha \geq k_H-1$. Let p be any point of H−{q}. Then $r_p \geq n+2$ so that p lies on two lines L and G of M. Lemma 11.4 a) shows that $(d_L-1)+(d_G-1) \geq 2\sqrt{(n-1)} - 4$. Hence

$$n^2-n-f+T' \leq n(n-1)- \sum_{L \in M} (d_L-1) \leq n(n-1)-(k_H-1)(2\sqrt{(n-1)} - 4)$$

so that

$$(+) \qquad (k_H-1)(2\sqrt{(n-1)} - 4)+T' \leq f \leq e-1.$$

It follows that $T' \leq k_H-1$ (if T' were at least k_H then we would obtain

$$2\sqrt{(n-1)} - 3 \leq (k_H-1)(2\sqrt{(n-1)} - 3) \leq e-2,$$

which contradicts $12 \leq 2n = e(e+1)$). In view of $T' \geq \alpha \geq k_H-1$, we obtain $T' = \alpha = k_H-1$. It follows that every point of H has degree n+2 and that every other point has degree n+1. Furthermore, (+) implies that $k_H = 2$, so $H = \{q,q'\}$ for a point q' outside of N. The lines L and G through q' miss N and satisfy $d_L+d_G+2 \geq 2\sqrt{(n-1)}$. In view of $4(n-1) = 2e(e+1)-4 > 2e^2$, we conclude that $d_L+d_G+2 \geq e\sqrt{2} > e+1 \geq f+2$. Hence $d_L+d_G \geq f+1$.

Assume that q' lies on a line X of degree n+1. Then q' lies on a unique line Y missing X. Since X has degree n+1, every point of Y has degree at least n+2, a contradiction, since every point $\neq$ q,q' has degree degree n+1.

Hence every line through q' has degree at most n. Since $d_L+d_G \geq f+1$ and because q' lies on the line H of degree two, it follows that $d_L+d_G = f+1$ and that every line $\neq$ H,G,L through q' has degree n (count the number of points using the lines through q').

Counting the number of points from q shows that q lies in a second line N' of degree n. N' meets n−1 of the n lines parallel to N. Hence N' meets L or G. Therefore q' lies also on a line F which misses N' and meets N. We have F $\neq$ H,G,L so F has degree n. But Lemma 11.7 says that N' meets every n-line. This contradiction completes the proof.■

<u>11.10 Lemma</u>. If there exists a line of degree n+1 then L is an extended L(n,e).

<u>Proof</u>. First consider the case that there exists a line L of degree n+1 which has a point q of degree at least n+2. In view of $r_p \geq n+1$ for every point p and because of $b = n^2+n+2$, this implies that q has degree n+2, that L meets every other line, and that every point other than q has degree n+1. It follows that every line of degree n+1 passes through q (if q were outside of a line L' of degree n+1 then q would lie on a line H parallel to L' and every point of H would have degree at least n+2). If p is any point other than q then the line X = pq satisfies $n^2+1-e \leq k_X+(r_p-1)(n-1)$. Hence $k_X \geq n+1-e > 2$ for every line X through q. We obtain therefore a linear space L' with $v' = v-1 \geq n^2-e$ points, $b' = b$ lines, constant point degree n+1, and maximal line degree n, if we remove the point q. By Lemma 11.2, L' is an L(n,e), so L is an extended L(n,e).

It remains to show that some line of degree n+1 contains a point of degree at least n+2. Assume to the contrary that every line of degree n+1 has only points of degree n+1, and fix any line L of degree n+1. Then L meets n^2+n other lines and is therefore parallel to a unique line H. Every point of H has degree n+2, every other point has degree n+1, and H is the unique line which is parallel to every line of degree n+1. Since H has no point of degree n+1, Lemma 11.9 shows that $t_N = 0$ for every n-line N. Thus H is also parallel to every n-line. Consequently, if q is any point of H then $n^2+1-e \le v \le k_H+(r_q-1)(n-2)$ so that $k_H \ge n+3-e$. Let p be any point outside of H. Since the lines px, $x \in H$, have degree at most n−1, we have

$$n^2+1-e \le v \le 1+k_H(n-2)+(r_p-k_H)n = n^2+n+1-2k_H \le n^2-n-5+2e.$$

It follows $3e \ge n+6$. Hence $n = e+3 = 6$ and $k_H = n+3-e = n$. But $t_N = 0$ for every n−line, a contradiction.■

<u>11.11 Lemma</u>. If every line has degree at most n then L is a closed L(n,e).

<u>Proof</u>. We prove this lemma in several steps. By Lemma 11.8, every point has degree at least n+1. We recall that we defined $d_L = n-k_L$ for every line L.

Step 1. *There is a point of degree* n+1.

For: If every point had degree at least n+2 then count incident point−line pairs to obtain the contradiction $bn \ge v(n+2)$.

Step 2. *If* p *is a point of degree* n+1 *and* L *is a line through* p, *then* $t_L-2 \le d_L \le f$ *and* $\sum_{p\in X} d_X = f$.

For: $\sum d_L = f$ follows from $n^2-f = v = 1+\sum(k_L-1)$ (sums running over the lines L through p). It follows that $d_L \le f$ for every line L passing through p and that p lies on at least $r_p-f > 1$ lines of degree n.

Now let L be any line through p and let $N \neq L$ be an n−line through p. Then L meets $k_L-1+t_L = n-1-d_L+t_L$ of the lines parallel to N. Since N is parallel to $n+1-t_N$ lines, we obtain $t_L \le d_L+2-t_N$.

Step 3. *If every line has degree at least* n−f *then* $T \ge f+2$ *with equality if and only if* $f = e-1$ *and if every line has degree* n *or* n−f.

For: By Lemma 11.2 and in view of $0 \le d_L \le f \le e-1$ for every line L, we have

$$\begin{aligned}0 \le \sum d_L(f-d_L) &= (2n-f-1)T-n[2n-f(f+1)]\\ &= (2n-f-1)(T-f-2)-(n-1)[2n-(f+1)(f+2)]\\ &\le (2n-f-1)(T-f-2)\end{aligned}$$

Hence $T \ge f+2$ with equality only if $d_L \in \{0,f\}$ for every line L and if $e(e+1) = 2n = (f+1)(f+2)$, which implies that $f = e-1$.

Step 4. *Every* n*-line has only points of degree* $n+1$.

For: Assume to the contrary that there is an n-line N with a point q of degree at least $n+2$. Then $t_N > 0$. Lemma 11.7 shows therefore that $t_N = 1$ and that every line parallel to N has degree at most $n-1$. Then Lemma 11.4 c) implies that $n^2-n-f+T-t_N \le (n+1-t_N)(n-1)$. Hence $T \le f+1$. Since every line has a point of degree $n+1$ (Lemma 11.9), Step 2 implies that every line has degree at least $n-f$. In view of $T \le f+1$, this contradicts Step 3.

Step 5. *Suppose it exists a line* H *which has no point of degree* $n+1$. *Then* $v+1-k_H \ge n^2+1-½n$ *and every point of H has degree at least* $n+3$.

For: Let p be any point of degree $n+1$ and let L be any line through p with $k_L < n$. Then Step 2 shows that p is contained in at most $1+(f-d_L)$ lines which have not degree n. Since H misses every n-line (this follows form Step 4, since H has no point of degree $n+1$), this implies that $k_H \le f+1-d_L \le f$.

Now let x be any point of H. Since q is not contained in any n-line (Step 4), we have $n^2-f = v \le k_H+(r_x-1)(n-2)$, so x has degree at least $n+3$.

Now consider two n-lines N and N' passing through the point p. The line N' meets $n-1$ of the $n+1$ lines parallel to N. Since H is parallel to N and N', it follows that H has a point q which does not lie on a second line parallel to N and N'. Because q has degree at least $n+3$, it is contained in two lines $X,Y \ne H$ which miss N. X and Y meet N' and have therefore a point of degree $n+1$ (Step 4). As before, it follows that $k_H \le f+1-d_Y$ and $k_H \le f+1-d_X$. From Lemma 11.4 a) we obtain $(d_X+1)(d_Y+1) \ge \quad n-t_N = n = ½e(e+1) \ge ½(f+1)(f+2)$. Since $d_Y+1 \le f+2-k_H \le f$ we conclude that $2d_X > f+1$ and $2k_H \le 2(f+1-d_X) \le f$. Now $v+1-k_H > n^2+1-½n$ follows from $v = n^2-f$, $4 \le 2k_H \le f$, and $2n = e(e+1) \ge (f+1)(f+2)$.

Step 6. *Every line has a point of degree* $n+1$; *in particular. every line has degree at least* $n-f$.

For: Assume to the contrary that there is a line H without a point of degree $n+1$ and denote by q a point of H. By Step 5, $r_q \ge n+3$.

We construct a new linear space L' from L in the following way. We remove the line H and the points $\ne q$ from H. If it happens that we produce lines of degree one in this way (these must be 2-lines of L which meet H in a point $\ne q$) then we remove also these lines (but not their points outside of H). The resulting structure L' is then in fact a linear space. It has $v+1-k_H > n^2+1-\frac{1}{2}n$ points (Step 5) and at most $b-1 = n^2+n+1$ lines. Furthermore, the point q has degree at least $n+2$ in L'. Theorem 7.4 shows that L' can be embedded into the closed complement D of a Baer-subplane in a projective plane of order n. Since D has a unique point q' of degree $\neq n+1$ which lies only in lines of degree $n+1-\sqrt{n}$, we conclude that q' corresponds to q and that every line $\neq$ H of L through q has degree at most $n+1-\sqrt{n}$.

In the same way follows that every line of L which meets H has degree at most $n+1-\sqrt{n}$. Hence if x is any point of degree $n+1$ then x lies on at least $k_H \geq 2$ lines which have degree at most $n+1-\sqrt{n}$. In view of $2(\sqrt{n}-1) > e-1 \geq f$, this contradicts Step 2.

Step 7. *If a line* L *satisfies* $t_L > d_L$ *then* $2d_L \geq f$ *and equality implies that* L *meets a line of degree* $n-f$.

For: Set $d = d_L$ and $t = t_L$ and suppose $t > d$. Assume by way of contradiction that every line parallel to L has degree at most $n-1$. Since L misses $m := (d+1)n+1-t \leq (d+1)n-d$ lines and because each of the $v-k_L$ points outside L is contained in at least $d+1$ of the lines parallel to L, we obtain

$$(n^2-f+d-n)(d+1) = (v-k_L)(d+1) \leq m(n-1) \leq [(d+1)n-d](n-1).$$

Hence $d(n+d-f) \leq f$. In view of $d \geq 1$ (Step 4), we obtain $n+1 \leq 2f$, which contradicts $2n = e(e+1) \geq (f+1)(f+2)$.

Consequently, L is parallel to an n-line N. In view of $t > 0$, L contains a point q of degree at least $n+2$. Let H be a second line through q which is parallel to N. Lemma 11.4 shows $2(d+1)(d_H+1) \geq 2(n-t_N) = 2n = e(e+1) \geq (f+1)(f+2)$. In view of $d_H \leq f$ (Step 6), we conclude $2d \geq f$ with equality only if H has degree $n-f$.

Step 8. $T \leq f+2$.

For: First consider the case that there is a line L of degree $n-f$ and let p be a point of degree $n+1$ of L. By Step 2, $t_L \leq f+2$ and L is the only line through p which has not degree n. In view of $t_N = 0$ for every n-line N, this implies $T = t_L \leq f+2$.

Now suppose that there is no line of degree $n-f$ and let p be any point of degree $n+1$. Then Step 7 shows that $2d_L > f$ for every line L through p which satisfies $t_L > d_L$. Step 2 implies therefore that p lies on at most one line L with $t_L > d_L$ and that $t_L \le d_L+2$ in this case. It follows therefore from Step 2 that $T = \sum_{p \in L} t_L \le 2 + \sum_{p \in L} d_L \le f+2$.

Step 9. L *is a closed* L(n,e).

For: Step 3, Step 6 and Step 8 imply that $T = f+2 = e+1$ and that every line has degree $n-f$ or n. It follows that every point of degree $n+1$ lies on a unique line of degree $n-f$ and on n lines of degree n.

In view of $T = f+2$, there is a point q of degree at least $n+2$. By Step 4, q lies only on lines of degree $n-f$. Hence $n^2-e = v-1 = r_q(n-f-1) = r_q(n-e)$ so that $r_q = n+e+2 = n+1+T$. It follows that every point other than q has degree $n+1$, so L is a closed L(n,e). This completes the proof of Lemma 11.11.■

Lemma 11.10 and Lemma 11.11 prove part a) of Theorem 11.1 and this completes the proof of this theorem.

We already pointed out that no example for an L(n,e) with $2n = e(e+1)$ is known. Therefore the following problem is interesting in the study of linear spaces with few lines (especially for those values of n which do not satisfy $2n = e(e+1)$ for any integer e).

<u>11.12 Problem</u>. Given an integer $n \ge 6$ what is the largest number of points of a linear space with n^2+n+2 lines which can not be embedded into a closed L(n,e) or an extended L(n,e).

12. Points of degree n and another characterization of the linear spaces L(n,d)

Corollary 4.8 says that every point of a linear space with n^2+n+1 lines has degree at least $n+1$ provided that $v \geq n^2-n+3$. This is not longer true for $v = n^2-n+2$. The linear spaces $\mathbf{E}_3$, $\mathbf{E}_4$, $\mathbf{E}_8$ and $\mathbf{E}_9$ are counterexamples. Probably these are all linear spaces with $v = n^2-n+2$, $b = n^2+n+1$, and a point of degree n. We shall not succeed to prove this, but we shall be able that describe very precisely every other example in this section. Each example is closely related to an L(n,d). Suppose an (n+1,1)-design D is an L(n,d) where $2n = d(d+1)$, i.e. every point lies on a unique line of degree $n-d$ and on n lines of degree n (see Definition 10.1 and Lemma 10.2). Then D has n^2+n+2 lines. Let N be an n-line of D, let Π be the parallel-class containing N, and let L be a line of degree $n-d$ which is not in Π. For $n \geq 6$, we obtain a linear space from D in the following way. We add an infinite point to the lines of Π and we remove the line L together with all but one of its points. The resulting linear space has $n-n+2$ points and n^2+n+1 lines. Furthermore, the point of L which we did not remove has degree n in L. Every linear space which can be obtained from an L(n,d) in this way, will be called a *reduced* L(n,d).

<u>12.1 Theorem</u>. Suppose L is a linear space with n^2-n+2 points and n^2+n+1 lines which has a point of degree at most n. Then L is one of the linear spaces $\mathbf{E}_3$, $\mathbf{E}_4$, $\mathbf{E}_8$, and $\mathbf{E}_9$, or $2n = d(d+1)$ for an integer d and L is a reduced L(n,d).

In section 10 we gave a non-existence criterion for an L(n,d). Together with this, Theorem 12.1 restricts the possible values of n for a linear space with n^2-n+2 points, n^2+n+1 lines, and a point of degree n very strongly (see Corollary 12.12). For example, if $n \neq 3,4$, then n is never the order of a classical projective plane. The first two possible values for n are 6, and 10 (in the next section we shall show that $n = 6$ does not occur). From this point of view, our result is an interesting one. But it is not a real classification, because we do not know about the existence of any L(n,d) with $2n = d(d+1)$.

We shall prove the theorem in several lemmata. Throughout, L denotes a linear space satisfying the hypothesis of 12.1. In view of Theorem 4.7 we may assume that L has a point q of degree n which lies on a line G of degree n+1 and on n−1 lines of degree n. Notice that this implies that every line $\neq$ G has degree at most n and that every point outside of G has degree at least n+1. Since every linear space with four points has four or six lines, we have $n \geq 3$.

<u>12.2 Lemma</u>. Every point of G has degree at least n with equality if and only if it lies on n−1 lines of degree n.

<u>Proof</u>. Let p be any point of G. Since every line $\neq$ G has degree at most n, we have $n^2-2n+1 = v-k_G \leq (r_p-1)(n-1)$. Thus $r_p \geq n$ with equality if and only if every line $\neq$ G through p has degree n.∎

<u>12.3 Lemma</u>. a) If G meets every line, then every point outside of G has degree n+1 and $\sum_{p\in G}(r_p-n) = n+1$.

b) If there is a line H parallel to G, then H is the only line parallel to G and $\sum_{p\in G}(r_p-n) = n$. In this case, every point outside of G and H has degree n+1, and every point of H has degree n+2.

<u>Proof</u>. a) Since G meets every other line, every point outside of G has degree $k_G = n+1$, and $\sum_{p\in G}(r_p-n) = b-|\{G\}|-k_G(n-1) = n+1$.

b) Let N and N' be two lines through q which intersect H. Then the points $p = N\cap H$ and $p' = N'\cap H$ have degree at least $k_G+1 = n+2$, q has degree n and every other point of $N\cup N'$ has degree at least $k_G = n+1$. Thus N intersects at least n^2 other lines and N' intersects at least n lines which are parallel to N. In view of $b = n^2+n+1$, we conclude that every line meets N or N'. Furthermore, p and p' have degree n+2 and every point of $N\cup N'-\{q,p,p'\}$ has degree n+1.

Let H' be any line parallel to G. We may assume that H' meets N. Since $N\cap H'$ has degree at least $k_G+1 = n+2$, we have $N\cap H' = q$. Since q has degree n+2, it lies only on one line missing G. Hence, H = H' and H is the only line parallel to G. It follows that every point of H has degree k_G+1 and that every point outside of G and H has degree $k_G = n+1$. It follows that $\sum_{p\in G}(r_p-n) = b-|\{G,H\}|-k_G(n-1) = n$.∎

Hence, G meets every line or G is parallel to a unique line. Each of these two cases will be studied seperately. We shall distinguish also the cases that q is the only point of degree n and that there are at least two points of degree n. This gives four possibilities. Two of them can already been handled very easily.

12.4 Lemma. If G meets every other line and if q is the only point of degree n, then there is an integer d with $2n = d(d+1)$ and L is a reduced L(n,d).

Proof. Since every point other than q of G has degree at least n+1, Lemma 12.3 a) shows that G has a unique point ∞ of degree n+2 and n−1 points of degree n+1. By Theorem 9.11, the incidence structure L' obtained from L by removing the point ∞ (but no line, even not the lines of degree 2 through ∞, if there are any) can be obtained from an L(n,d) with $2n = d(d+1)$ by removing a line L of degree n−d with all but one of its points. The set of lines of L through ∞ consist in L' of the line G, which has degree n in L', and the lines parallel to G. By definition, L is a reduced L(n,d).■

12.5 Lemma. If G is parallel to a line H and if q is the only point of degree n of L, then $L = E_4$.

Proof. Lemma 12.3 b) implies that every point $\neq q$ of G has degree n+1. Since the points of H have degree n+2, L can not be embedded into a projective plane of order n. Theorem 4.7 shows therefore that $L = E_4$.■

There remains the case that there are at least two points of degree n. We shall show that this implies that $n \leq 4$. In the proof we shall often consider sets consisting of the lines parallel to a given line. Informations about these sets can then be obtained from the following lemma.

12.6 Lemma. Let L be a line which intersects G in a point p, set $k_L = n+1-d$ and $r_p = n+1+f$, and let M be the set of lines which are parallel to L and which intersect G.

a) If G intersects every other line, then $|M| = dn-f$ and

$$\sum_{X \in M} k_X = d(n^2-2n+1+d)-f.$$

b) If G is parallel to a line H, then $|M| = dn-f-1$ and

$$\sum_{X\in M} k_X = \begin{cases} d(n^2-2n+1+d)-f-1, & \text{if } H\cap L = \emptyset \\ d(n^2-2n+1+d)-f+k_H-2, & \text{if } H\cap L \neq \emptyset. \end{cases}$$

<u>Proof</u>. a) L meets $(k_L-1)n+r_p-1$ other lines, so $|M| = dn-f$. Since every line of M meets G and because each of the $v+1-k_L-k_G$ points x outside $G\cup L$ lies on $r_x-k_L = d$ lines of M, it follows that

$$\sum_{X\in M} (k_X-1) = (v+1-k_L-k_G)d = (n^2-3n+d+1)d = d(n^2-2n+1+d)-f-|M|.$$

b) If L misses H, then L is parallel to $dn-f$ lines. In this case $|M| = dn-f-1$, since $H \notin M$. Furthermore, every point outside L and G lies on exactly d lines of M. If L meets H, then L is parallel to $dn-1-f$ lines and again we have $|M| = dn-1-f$. But this time, the k_H-1 points $\neq H\cap L$ of H lie on two lines of M. As in part a), this implies the equations of part b).■

<u>12.7 Lemma</u>. If G is parallel to a line H and if G has two points q and q' of degree n, then $n \leq 4$.

<u>Proof</u>. We shall prove this lemma in several steps. Set $d_X = n+1-k_X$ for every line X.

Step 1. H is the only line parallel to G, every point of H has degree $n+2$, and every point outside of G and H has degree $n+1$. Furthermore, $\sum_{p\in G} (r_p-n) = n$.

For: see Lemma 12.3 b).

Step 2. Let N be an n-line which passes through a point of degree n of G and which meets H, and set $\Pi = \{N\} \cup \{X \mid X$ is parallel to $N\}$.

a) $|\Pi| = n+1$, $k_X \leq n-1$ *for all* $X \in \Pi-\{N\}$, *and* $\sum_{X\in\Pi} k_X = n^2-n+k_H+1$.

b) Suppose X and Y are lines of Π which meet. Then $d_X d_Y \geq n$.

For: a) Since N contains a point of degree n, every line of $\Pi-\{N\}$ has degree at most $n-1$. This and Lemma 12.6 b) imply a).

b) This follows from the Parallel-Lemma in Section 1.

Step 3. Suppose that N_1 and N_2 are n-lines which meet G in points of degree n (not necessarily the same) and set $\Pi_j = \{N_j\} \cup \{X \mid X \neq H$ and X is parallel to $N_j\}$, $j = 1,2$. Then $|\Pi_1| = |\Pi_2| = n+1$ and $|\Pi_1\cap\Pi_2| \leq 2$.

For: Notice first that N_1 and N_2 meet, because they contain points of degree n. By Lemma 12.6, $|\Pi_j| = n+1$. Let p be any point $\neq N_2 \cap N_1, N_2 \cap G$ of N_2. If N_2 meets H and if $p = N_2 \cap H$, then p has degree n+2 and lies on two lines parallel to N_1. At least one of these two lines is different of H. In every other case, p has degree n+1 and lies on a unique line parallel to N_1. Thus, N_2 meets at least n−2 lines $\neq$ H which are parallel to N_1. Because N_2 meets also N_1, this shows that $|\Pi_1 \cap \Pi_2| \le |\Pi_1| - (n-1) \le 2$.

Step 4. Let N *be an* n*-line which meets* G *in a point of degree* n *and which meets* H. *Suppose that every line of degree* n−1 *which misses* N *contains a point of degree* $\neq$ n+1. *Then* $n \le 4$.

For: First notice that the point of intersection of G and a line $\neq$ H of degree n−1 has degree at least n+1, since an n-point of G lies only on G and n-lines. Set $\Pi = \{N\} \cup \{X \mid X \text{ is parallel to } N\}$.

Assume that $n \ge 5$. By Step 2 a), Π contains at least k_B+1 lines of degree n−1. Since $n \ge 5$, any two such lines are disjoint by Step 2b). Hence, there are two disjoint lines L and L' of degree n−1 in Π which miss H. Set $p = L \cap G$ and $p' = L' \cap G$. Since every point outside of G and H has degree n+1, the hypothesis in Step 4 shows that p and p' have degree at least n+2. It follows therefore from Step 1 that there are three points q_1, q_2 and q_3 of degree n on G. Let N_j be the line through q_j which is parallel to L, and set

$$\Pi_j = \{N_j\} \cup \{ X \mid X \neq H \text{ and } X \text{ is a line parallel to } N_j\}.$$

W.l.o.g. $N = N_1$ so that $\Pi_1 = \Pi$. Lemma 12.6 b) shows that $|\Pi_j| = n+1$. In view of $L \in \Pi_j$, Step 3 shows also that $|\Pi_1 \cup \Pi_2 \cup \Pi_3| \ge 3n-2$.

Let M be the set of lines $\neq$ H parallel to L. Because every point $\neq$ p of L has degree n+1, every line of $\Pi_j - \{L\}$ which meets L has to meet L in p. Hence, if M_0 the set of lines in $\Pi_1 \cup \Pi_2 \cup \Pi_3$ which do not contain p, then $M_0 \subseteq M$.

Set $r_p = n+1+f$. Then $|M| = 2n-1-f$ (Lemma 12.6 b)) and f+1 of the lines of Π_j pass through p. Since one of these f+1 lines is L, it follows that $|M_0| \ge |\Pi_1 \cup \Pi_2 \cup \Pi_3| - 3f \ge 3n-3f-3$, so $3n-3f-3 \le |M_0| \le |M| = 2n-1-f$, i.e. $n \le 2+2f$, which shows that $f \ge 2$ and $r_p = n+1+f \ge n+3$.

Consequently, p lies on at least 3 lines which miss $N_1 = N$. One of these lines is L. Let C and D be two others. Then $C, D \in \Pi = \Pi_1$. By Step 2 b), $2d_C = d_L d_C \ge n$

and $2d_D \geq n$. In the same way follows that $\mathbb{T}$ contains lines E and F which have degree at most $1+\frac{1}{2}n$ and which pass through $p' = L' \cap G$. Now, Step 2 a) shows that

$$n^2-n+1+k_H \leq k_N+k_C+k_D+k_E+k_F+(|\mathbb{T}|-5)(n-1)$$

$$\leq n+4(1+\frac{1}{2}n)+(n-4)(n-1) = n^2-2n+8.$$

It follows that $n \leq 7-k_H \leq 5$. In view of $n \geq 5$, we obtain equality in all the above inequalities. Thus $n = 5$ and C, D, E, and F have degree $1+\frac{1}{2}n$. But the degree of a line must be an integer. This contradiction completes the proof of Step 4.

Step 5. $n \leq 4$.

For: Assume to the contrary that $n \geq 5$. Let N be an n-line which meets H and passes through the point q of degree n, and set $\mathbb{T} = \{N\} \cup \{X \mid X \text{ misses } N\}$. By Step 4, $\mathbb{T}$ contains a line L of degree $n-1$ with only points of degree $n+1$. Let N' be the unique line of degree n which misses L and passes through the second point q' of degree n of G, and set $\mathbb{T}' = \{N'\} \cup \{X \mid X \text{ misses } N' \text{ and } X \neq H\}$. Then $L \in \mathbb{T} \cap \mathbb{T}'$, $|\mathbb{T}| = |\mathbb{T}'| = n+1$, $|\mathbb{T} \cap \mathbb{T}'| \leq 2$ (Step 3) and $|\mathbb{T} \cup \mathbb{T}'| = 2(n+1)-|\mathbb{T} \cap \mathbb{T}'| \geq 2n$. If $\mathcal{M}$ denotes the set of lines $\neq H$ which are parallel to L, then $|\mathcal{M}| = 2n-1$ (Lemma 12.6) and $\mathbb{T} \cup \mathbb{T}'$ is contained in $\mathcal{M} \cup \{L\}$, since every point of L has degree $n+1$. It follows that $\mathbb{T} \cup \mathbb{T}' = \mathcal{M} \cup \{L\}$ and $|\mathbb{T} \cap \mathbb{T}'| = 2$.

Since G has two points of degree n, Step 1 shows that G has a point z of degree at least $n+2$. Let $n+1+f$ denote the degree of z. Then z is contained in $f+1$ lines of $\mathbb{T}$ and in $f+1$ lines of $\mathbb{T}'$. These lines are also parallel to L. Since z is contained only in $f+2$ lines parallel to L, this shows that z lies on at least f lines of $\mathbb{T} \cap \mathbb{T}'$. In view of $L \in \mathbb{T} \cap \mathbb{T}'$ and $|\mathbb{T} \cap \mathbb{T}'| = 2$, this implies that z has degree $n+2$ and that z lies on the unique line Y of $\mathbb{T} \cap \mathbb{T}' - \{L\}$. This argument shows also that z is the only point of degree at least $n+2$ of G.

Assume by way of contradiction that Y has a point p of degree $n+1$. Then Y is the only line of $\mathbb{T}$ or $\mathbb{T}'$ through p. However, p lies on two lines parallel to L and $\mathcal{M} \cup \{L\} = \mathbb{T} \cup \mathbb{T}'$, a contradiction.

Consequently, every point of Y has degree at least $n+2$. Since every point of degree $\neq n+1$ lies in $H \cup G$, it follows that $k_Y = 2$ and that Y meets H in a point y. As N, also the line qy has degree n and meets H. It follows therefore in the same way that z lies on a line Y' of degree 2 which misses qy and meets H in a point y'. Because qy meets Y, we have $Y \neq Y'$.

Let N_1 be any n-line $\neq qy, qy'$ through q. Then N_1 is parallel to the two 2-lines Y and Y'. If M_1 is the set of lines $\neq H$ parallel to N_1, then it follows that the sum s of the degrees of the lines of M_1 is at most $|M_1-\{Y,Y'\}|(n-1)+4$. However $s \geq n^2-2n+2$ and $|M| = n-1$ by Lemma 12.6, a contradiction, since $n \geq 5$.

This contradiction completes the proof of Step 5 and the lemma.■

<u>12.8 Lemma</u>. Suppose that G intersects every other line and G has two points q and q' of degree n.

a) A line of degree n intersects G in a point of degree n or $n+1$.

b) It exists a line of degree $n-1$.

<u>Proof</u>. a) Assume to the contrary that G has a point p of degree $n+1+f \geq n+2$ which is contained in a line N of degree n. By Lemma 12.6, N is parallel to a line N' of degree n. Set $N' \cap G = p'$, denote the degree of p' by $n+1+f'$, and let L be a second line through p which is parallel to N'. Every point $x \neq p$ of L is contained in a line L_x which is parallel to N. Since L is the unique line through x which is parallel to N', we see that N' meets L_x. Since every point outside of $G \cup N$ lies on a unique line parallel to N, all the lines L_x pass through p'. Because p' lies on f' lines $\neq N$ which are parallel to N, there are at most f' different lines L_x, i.e. $k_L \leq f'+1$. Lemma 12.6 used for N' shows that $n^2-2n+2-f' \leq (n-1-f')n+k_L \leq (n-1-f')n+f'+1$. We obtain $f' \leq 1$ so that $2 = k_L \leq f'+1$ implies that $k_L = f'+1 = 2$. If w is the point $\neq p$ of L, then it follows now in the same way that L_w has degree 2. Hence

$$n^2-n+1 = v-1 = \sum_{w \in X} (k_X-1) \leq (r_w-2)(n-1)+1+1 = n^2-2n+3.$$

so that $n \leq 2$, which is a contradiction.

b) Assume to the contrary that there is no line of degree $n-1$. In view of $v > k_G+(n+1)(n-3)$, part a) implies then that G has no point of degree $n+2$.

Let N be a line of degree n through q, and let π be the set of lines parallel to N. Then $k_X \leq r_q-1 = n-1$ and thus $k_X \leq n-2$ for every line of π. If α denotes the number of lines of degree $n-2$ in π, then Lemma 12.6 a) shows that

$$n^2-2n+3 = \sum_{X\in \mathbb{T}} k_X \le (|\mathbb{T}|-\alpha)(n-3)+\alpha(n-2) = |\mathbb{T}|(n-3)+\alpha = n^2-2n-3+\alpha.$$

If follows that $\alpha \ge 6$. In view of $\alpha \le |\mathbb{T}| = n+1$, we obtain $n \ge 5$.

Suppose that $n = 5$. Then all the lines of $\mathbb{T}$ have degree $n-2 = 3$. In the same way follows that each line which is parallel to one of the lines passing through q has degree $n-2$. Hence, every line $\neq G$ has degree $n-2 = 3$ or degree $n = 5$. Thus, if β is the number of lines of degree 5 passing through a point x outside of G, then $21 = v-1 = \beta 4+(r_x-\beta)2 = 12+2\beta$, which is a contradiction.

If $n = 6$, then each of the 5 lines of degree n through q is parallel to a unique line of degree $n-3 = 3$. Since every 3-line is parallel to 3 lines through q, we obtain $3|5$, which is again a contradiction.

Consequently, $n \ge 7$. Let L be a line of degree $n-2$ of $\mathbb{T}$, set $p = L\cap G$ and $r_p = n+1+f$, denote by $\mathcal{M}$ the set of lines missing L, and let c be the number of lines of $\mathcal{M}$ which have degree at most $n-2$. Then Lemma 12.6 shows that $|\mathcal{M}| = 3n-f$ and

$$3(n^2-2n+4)-f = \sum_{X\in \mathcal{M}} k_X \le c(n-2)+(|\mathcal{M}|-c)n = |\mathcal{M}|n-2c.$$

It follows that $2c \le 6n-12+f-fn$.

Let $\mathbb{T}'$ be the set of lines of $\mathbb{T}$ which do not pass through p. Then $\mathbb{T}'$ consists of the lines missing N and L and $|\mathbb{T}'| = |\mathbb{T}|-f-1 = n-f$. Let q' be a second point of degree n of G, let N_1 and N_2 be the lines through q' which miss L, and denote by $\mathbb{T}'_j$ the set of lines which are parallel to N_j and L. Then $|\mathbb{T}_j| = n-f$. Set $\mathcal{M}' = \mathbb{T}' \cup \mathbb{T}'_1 \cup \mathbb{T}'_2$. Then $|\mathcal{M}'| \ge 3(n-f)-|\mathbb{T}'\cap\mathbb{T}'_1|-|\mathbb{T}'\cap\mathbb{T}'_2|-|\mathbb{T}'_1\cap\mathbb{T}'_2|$. The number of lines which miss N_1 and N_2 is two, and the number of lines which miss N and N_j is three (N and N_j meet, since N has an n-point). Because L misses N, N_1, and N_2 and is not in $\mathcal{M}'$, it follows that $|\mathcal{M}'| \ge 3(n-f)-2-2-1 = 3n-f-5$.

By definition, each line of $\mathcal{M}'$ is parallel to a line passing through the point q or q', so every line of $\mathcal{M}''$ has degree at most $n-1$. By hypotheses, every line of $\mathcal{M}'$ has therefore degree at most $n-2$. In view of $\mathcal{M}' \subseteq \mathcal{M}$ and because c is the number of lines of degree at most $n-2$ of $\mathcal{M}$, this shows that $|\mathcal{M}'| \le c$, i.e. $2(3n-f-5) \le 2c \le 6n-12+f-fn$. It follows that $2+f(n-7) \le 0$. In view of $n \ge 7$, this is the desired contradiction.■

<u>12.9 Lemma</u>. If G intersects every other line and if G has two points q and q' of degree n, then $n \leq 4$.

<u>Proof</u>. By 12.8 b), there is a line L of degree n−1. Set $p = L \cap G$ and $r_p = n+1+f$, let N be the unique line parallel to L through q and N' the unique line parallel to L' through q'. Notice that N and N'are not parallel. Furthermore define

$$\mathcal{M} = \{X \mid X \text{ is parallel to } L\},$$

$$\Pi = \{X \mid X \text{ is parallel to } N\}, \qquad \Pi_0 = \{X \in \Pi \mid p \notin L\},$$

$$\Pi' = \{X \mid X \text{ is parallel to } N'\}, \qquad \Pi'_0 = \{X \in \Pi' \mid p \notin L\},$$

$$\mathcal{M}' = \Pi_0 \cup \Pi'_0, \quad \mathcal{S} = \Pi_0 \cap \Pi'_0 = \{X \in \Pi \cap \Pi \mid p \notin X\}, \text{ and } \quad s := |\mathcal{S}|.$$

Lemma 12.6 a) shows that $|\mathcal{M}| = 2n-f$ and $|\Pi| = |\Pi'| = n+1$. It follows that $|\Pi_0| = |\Pi'_0| = n-f$, $|\Pi \cap \Pi'| = 3$, and $|\mathcal{M}'| = 2n-2f-s$. In view of $L \in \Pi \cap \Pi'$, there are lines Y and Z with $\Pi \cap \Pi' = \{L,Y,Z\}$. The set $\mathcal{S}$ is contained in $\{Y,Z\}$, so $s \leq 2$. Every line of $\mathcal{M}'$ has degree at most n−1 and $\mathcal{M}' \cup \{N,N'\}$ is contained in $\mathcal{M}$. Using Lemma 12.6 a) for L, we obtain

$$2(n^2-2n+3)-f = \sum_{X \in \mathcal{M}} k_X \leq |\mathcal{M}-\mathcal{M}'|n + |\mathcal{M}'-\mathcal{S}|(n-1) + \sum_{X \in \mathcal{S}} k_X$$

$$= (f+s)n + 2(n-f-s)(n-1) + \sum_{X \in \mathcal{S}} k_X.$$

It follows that

$$(+) \qquad f(n-3) \leq 2n-6+s-s(n-1) + \sum_{X \in \mathcal{S}} k_X \leq 2(n-3)+s.$$

First consider the case $s \leq 1$ and assume w.l.o.g. that $Y \notin \mathcal{S}$. Then Y contains p, so $Y \notin \mathcal{M}$. Let y be a point $\neq p$ of Y. Since y has degree n+1, Y is the unique line of Π and the unique line of Π' through y. The two lines of $\mathcal{M}$ through y are therefore not contained in $\{N,N'\} \cup \mathcal{M}'$. It follows that

$$2n-f = |\mathcal{M}| \geq 2+|\mathcal{M}' \cup \{N,N'\}| = 2n+4-2f-s \geq 2n+3-2f.$$

Hence, $f \geq 3$. Therefore (+) implies that $n \leq 4$.

Now consider the case $s = 2$. Then $\mathcal{S} = \{Y,Z\}$, $p \notin Y,Z$ and $|\mathcal{M}'| = 2n-2-2f$. Let y be any point of degree n+1 of Y. Then Y is the unique line of Π and the unique line of Π' through y. Since y lies on two lines of $\mathcal{M}$ (one of which is Y), y lies on one line of $\mathcal{M}-(\mathcal{M}' \cup \{N,N'\})$. Because Y has at least k_Y-1 points of degree n+1, we conclude that $2n-f = |M| \geq k_Y-1+|M'\cup\{N,N'\}| = 2n-1-2f+k_Y$, thus $f \geq k_Y-1 \geq 1$. In the same way follows that $f \geq k_Z-1$. Now (+) shows that $f(n-3) \leq k_Y+k_Z-2 \leq 2f$. Hence, $n \leq 5$. It remains to show that $n \neq 5$.

Assume to the contrary that $n = 5$. Then equality holds in all the above inequalities. Hence, $k_Y = k_Z = f+1$ and every other line of M' has degree $n-1$ (see how we got (+)). $\mathbb{T}$ contains the $n-f = 5-f$ lines of $\mathbb{T}_0 = \mathbb{T}\cap M'$, the line L, and f other lines through p. Set $\mathbb{T}_0-\{Y,Z\} = \{L_1,\ldots,L_{3-f}\}$ and let $H_1,\ldots,H_f$ be the lines $\neq$ L of $\mathbb{T}$ through p. Then L,Y,Z,H_j,L_j are the $n+1 = 6$ lines of $\mathbb{T}$. L and the lines L_j have degree $n-1 = 4$, and Y and Z have degree $f+1$. Lemma 12.6 a) shows now that

$$18 = n^2-2n+3 = \sum_{X\in\mathbb{T}} k_X = (4-f)4+2(f+1)+\sum_{j=1}^{f}|H_j| = 18-2f+\sum_{j=1}^{f}|H_j|.$$

Thus the lines H_j have degree 2. In the same way follows that p is contained in f lines of degree 2 of $\mathbb{T}'$. Since the lines of $\mathbb{T}\cap\mathbb{T}' = \{L,Y,Z\}$ do not contain p, p is therefore contained in 2f lines of degree 2. Counting the number of points from p, we see

$$16 = v-k_G = \sum_{p\in X\neq G}(k_X-1) \leq (k_L-1)+2f\cdot 1+(r_p-2-2f)(n-1)$$

$$= k_L-1+2f+(n-1-f)(n-1) = 19-2f.$$

Consequently $f \leq 1$. In view of $f = k_Y-1 \geq 1$, we obtain $f = 1$ and $k_Y = 2$. Let y be the point $\neq G\cap Y$ of Y. Then $v = n^2-n+2$ yields that the n lines $\neq$ Y through y have degree n. In view of $p \notin Y$, we conclude that py has degree n, which contradicts Lemma 12.8 a), since p has degree $n+1+f = n+2$.■

<u>Proof of Theorem 12.1</u>. In view of Lemma 12.4 and Lemma 12.5, we may assume that G has two points q and q' of degree n. Lemma 12.7 and Lemma 12.9 show therefore that $n \leq 4$. Since $v < b$, the linear space L is not a near pencil, so $v \geq k_G+2$, which implies that $n \geq 3$.

Consider first the case $n = 3$. Then there are $n^2+n+1 = 13$ lines and five of them contain q or q'. Let T be the set of the other 8 lines. In view of $v-k_G = 4$, every point of G has degree at most 5.

Assume by way of contradiction that G meets every other line. Then each of the 8 lines of T contains one of the two points $\neq$ q,q' of G. Thus each of the two points $\neq q,q'$ of G has degree 5 and is contained in 4 lines of degree 2. Let N_1,N_2 be the two 3-lines through q, let L_1,L_2 the two 3-lines through q', and let L the line which contains $N_1 \cap L_1$ and $N_2 \cap L_2$. Then L meets G in a point $p \neq q,q'$. But p is not contained in a 3-line, a contradiction.

Consequently, G is parallel to a line H. Since q has degree 3, the line H has degree 2. Now, one of the points $\neq q,q'$ of G has degree 4 and is contained in a 3-line and in two 2-lines and the other one has degree 5 and is contained in four 2-lines. Hence, $L = \mathbf{E}_8$ in this case.

Consider finally the case $n = 4$. then every line $\neq G$ has degree 2,3, or 4. If b_k denotes the number of lines of degree k, then $b_5 = 1$,

$$\sum b_k = 21$$

$$\sum b_k k(k-1) = v(v-1) = 182.$$

We have another equation: $\sum b_k k = \sum r_p$. However, we do not know the degree of all points. But we can do the following. Set $h = 0$, if G intersects every line and $h = k_H - 1$, if G is parallel to a line H. Then Lemma 12.3 shows that

$$\sum b_k k = \sum r_p = v(n+1)+h = 70+h.$$

In view of $b_5 = 1$, we obtain $b_2 = 6-3h$, $b_3 = 3+5h$, and $b_4 = 11-2h$. If a point p of G lies on t_k lines of degree k, then call (t_2,t_3,t_4) the *type* of p. We have $9 = v-k_G = t_2+2t_3+3t_4$ and $r_p - 1 = t_2+t_3+t_4$.

Assume by way of contradiction that G meets every other line so that $h = 0$. Lemma 12.8 a) implies that a 4-line meets G in a point of degree 4 or 5. Hence, the possible types for the points of G are (0,0,3), (0,3,1), (1,1,2), (1,4,0), (3,3,0) and (7,1,0). Since the points q and q' have type (0,0,3) and in view of $b_2 = 6$, $b_3 = 3$, and $b_4 = 11$, this is not possible.

Hence, G is parallel to a line H. By definition, H has degree h+1. Since H is parallel to the line G passing through the point q of degree $n = 4$, we have $k_H \leq 3$. In view of $v = 14$, each of the (at least six) points of outside of G and H, which have degree 5, lies in at least three 4-lines. Since every 4-line has 3 points outside of G, it follows that there are at least two 4-lines which do not contain q or q'. Hence $b_4 \geq 8$, which implies $h \leq 1$.

Consequently, $h = 1$, $b_2 = 3$, $b_3 = 8$, $b_4 = 9$, and H has degree 2. Since the possible types for the points of G are (0,0,3), (0,3,1), (1,1,2), (2,2,1) and (1,4,0) and because q and q' have type (0,0,3), it follows that the types of the three points of G−{q,q'} are (0,0,3), (1,4,0), (1,4,0), or (0,3,1), (0,3,1), (2,2,1).

Let D be the structure induced by L on the nine points outside of G. Then any two points of D are joined by a unique line (D is almost a linear space, it has just two lines of degree 1). It is easy to see that such a structure can not have four mutually disjoint parallel classes, three of which consists of three 3-lines and one of which consists of one 1-line and four 2-lines. This implies that it is not possible that three points of G have type (0,0,3) and one point of G has type (1,4,0).

Hence, G has one point of type (2,2,1), two points of type (0,3,1), and the two points q and q' of type (0,0,3). Let p_1, p_2, and p_3 be the three points $\neq$ q,q' of G, and let N_j be the unique n-line though p_j, $j = 1,2,3$.

Assume by way of contradiction that N_1 and N_2 intersect in a point p. Then p is contained in the 4-lines N_1, N_2, pq, and pq'. Together they cover $13 = v-1$ points. It follows that p has degree 5 and that the fifth line X through p has degree 2. X meets G and the point of intersection must be the point of type (2,2,1). Since p has degree 5, N_3 meets N_1 or N_2. We may assume that N_1 and N_3 meet in a point p'. As before, it follows that p' has degree 5 and that p' lies on a 2-line X' which meets G in a point of type (2,2,1). But G has only one point of this type, a contradiction.

Hence, N_1 and N_2 are parallel. In the same way follows that any two of the lines N_j are parallel. Consequently, each of the nine points outside of G lies on exactly one of the lines N_j and two of the six 4-lines of $\pi \cup \pi'$. Since the points of H have degree 6, it follows that they are contained in H, in three 4-lines, one 3-line, and one 2-line $\neq$ H. Thus, if p is the point of type (2,2,1) of G, then the two 2-lines trough p meet H, i.e. the three 2-lines of L form a triangle. If we replace the three 2-lines by the 3-line H∪{p}, we obtain therefore again a linear space L'. It has $n^2-n+2 = 14$ points and $n^2+n-1 = 19$ lines. Theorem 5.5 shows that L' can be embedded into a projective plane P of order 4. Since L' has still the two points q and q' of degree $n = 4$, there are lines L and L' of P−L' passing through q and q' respectively. L' arises from P by removing L and L' and the points $\neq q, q'$ from $L \cup L'$, and L arises from L' by breaking up a line of degree three into three lines of degree two. Thus $L = E_9$, and the proof of Theorem 12.1 is completed.■

12.10 Corollary. Suppose it exists a linear space L with n^2+n+1 lines, at least n^2-n+2 points, and a point of degree at most n. Then $v = n^2-n+2$ and one of the following case occurs.

1) L is one of the exceptional spaces E_3, E_4, E_8, and E_9.

2) There is an integer d with $2n = d(d+1)$ and L is a reduced L(n,d). Furthermore, d satisfies the conditions a), b), and c) of 10.8.

Proof. Corollary 4.8 yields $v = n^2-n+2$. Theorem 12.1 shows therefore that a) occurs or that L is a reduced L(n,d) with $2n = d(d+1)$. Lemma 10.4 implies d satisfies the conditions of 10.8.■

Corollary 12.10 and the results obtained in section 4 classify all linear spaces with at least n^2-n+2 points, at most n^2+n+1 lines, and with a point of degree at most n.

13. The Non-Existence of certain (7,1)-designs and determination of A(5) and A(6)

The smallest number n for which there is no projective plane of order n is six. This was already shown by Tarry (1900). It is therefore particularly interesting to ask for the maximal possible number v_{max} of points of a (7,1)-design with 43 lines. S. Vanstone (1973) showed that $v_{max} \leq 35$. In order to exclude the case $v_{max} = 35$, one has to show that there is no pseudo-complement of a hyperoval in a projective plane of order 6. This was done by P. de Witte (1977a) using a result on Graph-Theory of C. Li-Chien (1960). In several steps it was then shown that $v_{max} \in \{25,26,27\}$ (McCarthy 1976, McCarthy/Mullin 1976, McCarthy/Mullin/Schellenberg/Stanton/Vanstone 1976a/b, Mullin/Vanstone 1975, 1976b, and Metsch 1990a).

In this section, we shall give a common proof for the non-existence of the projective plane of order 6 and the pseudo-complement of a hyperoval in a projective plane of order 6. Then we show that every (7,1)-design has with 43 lines has at most 29 points. As an application, we prove that A(6) = 31 (see the definition following Corollary 7.5). We shall also prove that A(5) = 21.

The technique we use to show that every (7,1)-design with 43 lines has at most 29 points is the following. First we study (7,1)-designs D satisfying the following property.

(P) It exists a point q_0 and four 7-lines G_1, G_2, G_3 and H such that $q_0 \in G_1, G_2, G_3$ and $q_0 \notin H$. Furthermore, the number of lines is b = 43.

First we show that there is a unique (7,1)-design which satisfies (P) and has $v = |G_1 \cup G_2 \cup G_3 \cup H| = 23$ points, and that this design can not be extended (i.e. is not embedded in a (7,1)-design with more points). Then we shall see that every (7,1)-design with $b \leq 43$ lines has at most 29 points by showing that each counterexample contains a subdesign satisfying property (P).

Suppose that a (7,1)-design D with b = 43 lines and v = 23 points satisfies (P). The crucial point in the proof that D is uniquely determined is to show that the following condition is satisfied.

(C) If $X_j \neq H, G_j$ is a line through $H \cap G_j$, j = 1,2,3, and if X_3 meets X_1 and X_2, then X_1 meets X_2.

Then for any lines $X_2 \neq H,G_2$ through q_2 and $X_3 \neq H,G_3$ through q_3 there is a unique line $X_1 \neq H,G_1$ through q_1 such that X_1 meets X_2 and X_3, i.e. the lines X_j form a triangle. In this way we obtain 5 triangles $T_1,\dots,T_5$ and each of the 15 points $\neq q_0,\dots,q_3$ of $G_1 \cup G_2 \cup G_3$ is a point of exactly one of these triangles.

We obtain a linear space L, if we remove the points q_j, $j = 0,\dots,4$ and the four 2-lines of size through q_0. In L, each triangle T_j consists of three 2-lines. Hence we obtain a new linear space $\mathbf{L_1}$, if we replace the lines of T_j, $j = 1,\dots,5$, by a 3-line H_j. The set $\pi = \{H,H_1,\dots,H_5\}$ is a parallel class of $\mathbf{L_1}$. We let the lines of π intersect in a new point q and obtain in this way a linear space $\mathbf{L_2}$ with 20 points and 29 lines. The lines G_1, G_2, G_3, and H have size 5 and are mutually parallel in $\mathbf{L_2}$. Hence, $\mathbf{L_2}$ a pseudo-complement of two lines in a projective plane of order 5. Lemma 5.2 shows that $\mathbf{L_2}$ can be embedded, i.e. $\mathbf{L_2}$ is the complement of two lines in a projective plane P of order 5. The construction of P from D is clearly reversible, which shows that there is a (7,1)-design satisfying (P) and (C). Furthermore, the uniqueness of the projective plane of order 5, which is P(2,5), and its properties show that D is uniquely determined and that it can no be extended to a (7,1)-deisgn with more points (i.e. it has no parallel-class consisting of 7 lines).

<u>13.1 Theorem</u>. There is a unique (7,1)-design satisfying property (P). It has 23 points and it can not be extended.

For the proof of Theorem 13.1 it remains to show that a (7,1)-design with 23 points which satisfies (P) satisfies also (C).

<u>13.2 Lemma</u>. Suppose D is a (7,1)-design with 23 points which satisfies Property (P). Then D satisfies also Property (C).

<u>Proof</u>. By (P), there is a point q_0 which lies on three 7-lines G_1,G_2,G_3 and there is a 7-line H with $q \notin H$. Set $q_j = G_j \cap H$, let $p_1,\dots,p_4$ be the other four points of H, and denote by π_j, $j = 1,\dots,4$, the set consisting of the five lines $\neq H,qp_j$ through p_j. We set $\pi_1 = \{A_1,\dots,A_5\}$, $\pi_2 = \{B_1,\dots,B_5\}$, $\pi_3 = \{C_1,\dots,C_5\}$, and $\pi_4 = \{L_1,\dots,L_5\}$. Furthermore, denote by D' the incidence structure which is induced by the lines G_1, G_2, G_3, and the 20 lines of $\pi_1 \cup \pi_2 \cup \pi_3 \cup \pi_4$ on the 15 points $p_{ij} := G_i \cap L_j$ (which are

the points other than $q_0,\ldots,q_3$ of $G_1 \cup G_2 \cup G_3$). Then **D'** consists of the four parallel-classes $\{G_1,G_2,G_3\}$ and Π_j, $j = 1,\ldots,4$. It is easily verified that **D** satisfies (C) if and only if **D'** satisfies the following property.

(C') If x_j, $j = 1,2,3$, is a point of G_j and if x_1 is not joined to x_2 and x_3, then also x_2 and x_3 are not joined.

It suffices therefore to show that **D'** satisfies property (C'). Assume to the contrary that **D'** does not satisfy (C'). We denote by S_{ij} the set consisting of the three lines of $\Pi_1 \cup \Pi_2 \cup \Pi_3$ passing through p_{ij}. Since $p_{j1} \in L_1$, the sets S_{j1} are disjoint, so we may assume that $S_{j1} = \{A_j,B_j,C_j\}$, $j = 1,2,3$. Since (C') does not hold, we may furthermore assume that p_{12} is not joined to p_{21} and not joined to p_{31}, which means that $S_{12} \cap (S_{21} \cup S_{31}) = \emptyset$. Since $S_{11} \cap S_{12} = \emptyset$, we may therefore assume also that $S_{12} = \{A_4,B_4,C_4\}$.

A_1 B_1 C_1	A_4 B_4 C_4			
A_2 B_2 C_2	A_1 B_3 C_5	A_4	B_4	C_4
A_3 B_3 C_3		C_4	A_4	B_4

For convenience, we consider a diagram consisting of three rows of 5 squares, and identify the set S_{jk} with the k-th square of the j-th row. Thus the rows correspond to the lines G_j, $j = 1,2,3$, and the columns correspond to the lines L_j, $j = 1,\ldots,5$. Using that **D** is a (7,1)-design, it follows that

(P1) each square contains a unique element of each set Π_j, $j = 1,2,3$.

(P2) each element of $\Pi_1 \cup \Pi_2 \cup \Pi_3$ occurs exactly one time in each row.

(P3) each element of $\Pi_1 \cup \Pi_2 \cup \Pi_3$ occurs exactly at most one time in each column.

(P4) If $S_{ij} \cap S_{mn} \neq \emptyset$, then $|S_{ij} \cap S_{mn}| = 1$ and S_{ij} and S_{mn} are squares of different rows and columns.

Each element of S_{12} occurs one time in the second and third row but not in the first square (since we already know the sets S_{j1}) or the second square (P3). In view of (P3) and (P4), it is therefore no loss of generality to assume that $A_4 \in S_{23},S_{34}$, $B_4 \in S_{24},S_{35}$, and $C_4 \in S_{25},S_{33}$.

We have $|S_{22} \cup S_{32}| = 6$ and $A_4, B_4, C_4 \notin S_{22} \cup S_{32}$. Thus $3 \le |(S_{22} \cup S_{32}) \cap (S_{11} \cup S_{21} \cup S_{31})|$. We may assume that $2 \le |S_{22} \cap (S_{11} \cup S_{21} \cup S_{31})| = |S_{22} \cap (S_{11} \cup S_{31})|$. Since $|S_{22} \cap S_{11}| \le 1$ and $|S_{22} \cap S_{31}| \le 1$, we may assume $A_1, B_3 \in S_{22}$. Using the properties (P1), (P2), and (P3), it follows that $C_5 \in S_{22}$.

Each element of S_{31} occurs in a unique square in the first row which is not S_{11} or S_{12} (P2). Let S, S', and S" be squares in the first row with $A_3 \in S$, $B_3 \in S'$, and $C_3 \in S''$. Since $S_{31} = \{3,8,13\}$, (P4) shows that the squares S, S', and S" are the three last squares of the first row, in particular they are distinct.

We have $C_5 \notin S''$ (see (P1)) and $C_5 \notin S'$ (by (P4), since $B_3 \in S_{22} \cap S'$). Hence $C_5 \in S$ (by (P2)). Now we know the contribution of C_1, C_3, C_4, and C_5 in the first row, so that $C_2 \in S'$ (P4). We have $A_2 \notin S$ (since $A_3 \in S \cap M_1$) and $A_2 \notin S'$ (since $C_2 \in S' \cap S_{21}$). Thus $A_2 \in S''$. Now we know the contribution of A_1, A_2, A_3, and A_4 in the first row. It follows that $A_5 \in S'$. Finally, since S' and S" have already C_2 resp. A_2 in common with S_{21}, we have $B_2 \in S$, so that $B_5 \in S''$.

We have shown that $S = \{A_3, B_2, C_5\}$, $S' = \{A_5, B_3, C_2\}$, and $S'' = \{A_2, B_5, C_3\}$. In view of $\{S, S', S''\} = \{S_{13}, S_{14}, S_{15}\}$, there are six possibilities now. We shall obtain a contradiction in each case.

Suppose first that $S'' = S_{13}$. Then the only place for C_3 in the second row is the fourth square and the only place for B_5 in the second row is the fifth square. Now B_1 and C_1 must lie in the third square of the second row, a contradiction, since $B_1, C_1 \in S_{11}$.

From the other four possibilities, the easiest one is $S = S_{13}$, $S' = S_{14}$, and $S'' = S_{15}$.

A_1 B_1 C_1	A_4 B_4 C_4	A_3 B_2 C_5	A_5 B_3 C_2	A_2 B_5 C_3
A_2 B_2 C_2	A_1 B_3 C_5	A_4	B_4	C_4
A_3 B_3 C_3		C_4	A_4	B_4

In this case we obtain a contradiction in the following way. Since $B_5 \notin S_{21}, S_{22}, S_{24}$ (P1) and $B_5 \notin S_{25}$ (P3), we have $B_5 \in S_{23}$. Since C_3 lies not in S_{21} or S_{22} or S_{25} (P1), or S_{23} (this follows from (P4), since $B_5, C_3 \in S_{15}$ and $B_5 \in S_{23}$), we have $C_3 \in S_{24}$. Using the same arguments, it follows first that $A_5 \in S_{25}$ and then that $A_3 \in S_{24}$. But now, we have $A_3, C_3 \in S_{24} \cap S_{31}$, which contradicts (P4).

Next consider the case $S = S_{13}$, $S' = S_{14}$, and $S'' = S_{15}$. Then we have $C_3 \in S_{23}$, $B_5 \in S_{25}$ (since $C_3 \in S_{23} \cap S_{14}$), $C_1 \in S_{24}$, $B_1 \in S_{23}$, $A_5 \in S_{24}$, and $A_3 \in S_{25}$.

A_1 B_1 C_1	A_4 B_4 C_4	A_3 B_2 C_5	A_2 B_5 C_3	A_5 B_3 C_2
A_2 B_2 C_2	A_1 B_3 C_5	A_4 B_1 C_3	A_5 B_4 C_1	A_3 B_5 C_4
A_3 B_3 C_3		C_4	A_4	B_4

Now we obtain the contradiction only in the third row. First we have $C_1 \in S_{32}$ (Since $C_1 \notin S_{34}$ by (P3) and $C_1 \notin S_{35}$ (P4)). Then $C_2 \in S_{34}$, $C_5 \in S_{35}$, $A_5 \in S_{33}$ ($A_5 \notin S_{32}$, since $A_5, C_1 \in S_{24}$, and $A_5 \notin S_{35}$ by (P3)), $A_1 \in S_{35}$, and $A_2 \in S_{32}$. Now there is no more place for B_5 in the third row, because A_2 and C_4 already occur together with B_5 in a square.

In the case $S' = S_{13}$, $S = S_{14}$, and $S'' = S_{15}$, we obtain in the following order $B_5 \in S_{23}$, $B_1 \in S_{25}$, $C_3 \in S_{24}$, $A_3 \in S_{25}$, $C_1 \in S_{23}$, $A_5 \in S_{24}$, $C_5 \in S_{35}$, $A_1 \in S_{33}$ ($A_1 \notin S_{35}$, since $A_1, C_5 \in S_{22}$), $A_2 \in S_{32}$, $A_5 \in S_{35}$, $B_5 \in S_{34}$ ($B_5 \notin S_{32}$, since $A_2, B_5 \in S_{22}$), $C_1 \in S_{32}$ ($C_1 \notin S_{34}$, since $A_1, B_5 \in S_{23}$), $C_2 \in S_{34}$, and then there is no more place for B_1 in the third row.

In the case $S' = S_{13}$, $S'' = S_{14}$, and $S = S_{15}$, we obtain in the following order $C_3 \in S_{23}$, $B_5 \in S_{25}$, $B_1 \in S_{23}$, $A_3 \in S_{24}$, $A_5 \in S_{25}$, $C_1 \in S_{24}$, $C_1 \in S_{32}$, and then there is no place for B_1 in the third row. This completes the proof of the lemma.■

<u>13.3 Corollary</u>. There is no projective plane of order 6 and no pseudo-complement of a hyperoval in a projective plane of order 6.■

<u>13.4 Corollary</u>. If v is the number of points and if b_7 is the number of 7-lines of a (7,1)-design D, then $7b_7 \leq 2v$ or $v = 23$.

<u>Proof</u>. We may assume that $7b_7 > 2v$. Then there is a point q which lies on at least 3 lines of degree 7. If we can show that there exists a 7-line not passing through q, then Theorem 13.1 proves that $v = 23$. Assume to the contrary that q lies on every 7-line. Then $v-1 \geq 6b_7$. Together with $7b_7 > 2v$, we conclude that $v < b_7-1$. But $v \geq 7$, since there is a 7-line, and $b_7 \leq 7$, since every 7-line passes through q.■

<u>13.5 Theorem</u>. Every (7,1)-design with 43 lines has at most 29 points.

Proof. Let D be a counterexample with the maximal possible number v of points. Denote the number of k-lines by b_k. Lemma 3.4 shows that $b_6 = 0$. In view of $v > 29 = 1+7(5-1)$, every point lies on a 7-line. Hence $7b_7 \geq v$ so that $b_7 \geq 5$. Since a 7-line meets every other line, it follows that there is no line of degree 0.

Assume that a point p lies on three 7-lines. Then Theorem 13.1 shows that every 7-lines passes through q. Since every point lies on a 7-line, it follows that the 7-lines pass through q and cover every point. Hence $v \geq 1+b_7(7-1)$, so 6 divides v−1. If p is a point $\neq$ q, then every line $\neq$ pq through p has degree at most 5. Hence $v \leq 7+6(5-1) = 31$. Since $v \geq 29$ and because 6 divides n, it follows that v = 31 and that every line not passing through q has degree 5. Furthermore, q lies on five 7-lines (the other two lines through q must have degree 1). Hence D is the pseudo-complement of two lines in a projective plane of order 6 (if we forget the two lines of degree 1). Lemma 4.4 shows that D can be embedded. But there is no projective plane of order 6, a contradiction.

Consequently every point lies in at most two 7-lines. It follows that $v \leq 1+2(7-1)+5(5-1) = 33$. Furthermore, since a 7-line meets every other line, and because of $b_7 \geq 5$, it follows that there is no line of degree 1 or 2. The same argument shows that $b_7 \leq 6$, if there is a line of degree 3.

Assume that every point lies on two 7-lines. Then $7b_7 \geq 2v$. It follows that $7b_7 = 2v$ (Corollary 13.4). But $29 \leq v \leq 33$ so 7 does not divide v. Hence, there is a point which lies only on one line of degree 7. It follows that $v \leq 7+6(5-1) = 31$.

We have $\sum b_j = 43$, $\sum b_j \cdot j = v \cdot 7$, and $\sum b_j \cdot j(j-1) = v(v-1)$, which implies that $8b_3+3b_4 = \sum b_j(7-j)(5-j) = (43-v)(35-v)$, since $b_0 = b_1 = b_2 = 0$.

First assume that v = 31. Then $8b_3+3b_4 = 48$. Hence (b_3,b_4) is (6,0), (3,8) or (0,16). The equations $b_3+b_4+b_5+b_7 = 43$ and $3b_3+4b_4+5b_5+7b_7 = 7v$ show therefore that (b_3,b_4,b_5,b_7) equals to (6,0,30,7), (3,8,24,8) or (0,16,18,9). Since we have shown that $b_7 \leq 6$, if there is a 3-line, the fist two possibilities can not occur. Since $7b_7 \leq 2v$, the last possibility can also not occur.

Now consider the case v = 30. Then $8b_3+3b_4 = 65$. Since $v \equiv 0 \pmod 2$, every point must lie on an odd number of lines of even degree so that $2b_2+4b_4 \geq v$. It follows that (b_3,b_4) equals to (4,11) or (1,19) so that (b_3,b_4,b_5,b_7) equals to (4,11,21,7) or (1,19,15,8). But $b_7 \leq 6$, if there is a 3-line. Hence $v \leq 29$, which proves the lemma.■

<u>13.6 Corollary</u>. There does not exist an L(6,3).

<u>Proof</u>. By definition, an L(6,3) is a (7,1)-design with 33 points and 44 lines such that every point lines on one 3-line and six 6-lines (see definition 10.1 and Lemma 10.2). Hence, if we remove a 3-line, we obtain a (7,1)-design with 43 lines and 30 points. The preceding theorem shows that such a design does not exist.■

<u>13.7 Corollary</u>. There does not exist a (7,1)-design satisfying $30 < v < b \le 43$.

<u>Proof</u>. This follows from Lemma 3.6 and the theorem.

We defined A(n) to be the least integer ≥ 3 such that every non-degenerate linear space with $A(n) < v \le b \le n^2+n+1$ can be embedded into a projective plane of order n, and we proved that $A(p^{2s}) = p^{2s}-p^s+1$ for every prime power p^{2s}. This gave us A(4) = 15. We also know that A(2) = 3 and A(3) = 8. Since a projective plane of n−1 can not be embedded into a projective plane of order n, we have $A(n) \ge n^2-n+1$ whenever n−1 is the order of a projective plane. This shows that $A(5) \ge 21$ and $A(6) \ge 31$. We shall show now that equality holds. First we consider the case n = 6.

<u>13.8 Theorem</u>. It does not exist a non-degenerate linear space with $v \ge 32$ points and $b \le 43$ lines. In particular A(6) = 31.

<u>Proof</u>. Assume to the contrary that L is a non-degenerate linear space with $v \ge 32$ points and $b \le 43$ lines. Since there is no projective plane of order 6, Corollary 7.5 shows that $b = 43 = n^2+n+1$ with n := 6. Then Corollary 12.10 and 13.6 show that every point has degree at least n+1 = 7.

If we remove all points of degree ≠ 7, then we obtain a (7,1)-design. Corollary 13.5 shows therefore that there are at least two points q and q' which have degree ≠ 7. In Lemma 7.3 we have shown that

(1) every line has degree at most n,

(2) there exists a point of degree n+1 = 7 and every point of degree n+1 lies in at least three n-lines,

(3) an n-line has only points of degree n+1, and

(4) If an n-line is parallel to two intersecting lines H and L, then H or L has degree at most n−2.

We set $t := \sum (r_p - 7)$ where the sum runs over all points p. Let N be a line of degree $n := 6$ and denote by $\mathcal{M}$ the set of lines which are parallel to N. Because an n-line has only points of degree $n+1$, we have

$$(+) \qquad 26+t \leq v-k_N+t = \sum_{p \notin N} (r_p - k_N) = \sum_{X \in \mathcal{M}} k_X.$$

First consider the case that q and q' have degree at least 9. Then $t \geq 4$, and at least 5 lines of $\mathcal{M}$ contain q or q'. By (3) and (4), at most two of them have degree $n-1 = 5$ and the other three have degree at most 4. This contradicts (+).

We may therefore assume that q has degree 8. By (4), q lies on a line $L \in \mathcal{M}$ with $k_L \leq n-2 = 4$. In view of $v \geq 32$, it follows that $v = 32$, $k_L = 4$ and that the other seven lines through q have degree 5. Since every line which meets L in a point $\neq q$ is parallel to at least two of the 5-lines through q, (4) implies that L is parallel to every n-line. Hence, every point of L has degree at least $n+2$ (see (2)). It follows that $t \geq k_L = 4$. We have $L \in \mathcal{M}$, and L meets $k_L = 4$ other lines of $\mathcal{M}$, which have degree at most 5. Now (+) shows that $30 \leq v-6+t \leq k_L+4 \cdot 5+6 = 30$. Hence, every point of L has degree 8 and N is parallel to a unique line of degree n. Furthermore, $t = 4$, i.e. every point of L has degree $n+2$ and every other point has degree $n+1$. It follows as before that every point of L lies on seven lines of degree 5. Thus, if b_k is the number of k-lines, then $b_4 \geq 1$ and $b_5 \geq 28$. The equations $v(v-1) = \sum b_k \cdot k(k-1)$ and $b = 43$ imply therefore that $b_4 = 1$, $b_5 = 28$ and $b_7 = 14$. Let D be the (7,1)-design which is obtained by removing L together with its four points. Then D has 14 lines of degree $n = 6$, and each of these lines B misses a unique line of degree 6 and 4 lines of degree 4 (these are the four lines which miss B and meet L in L). Each line of degree 6 together with the 5 lines parallel to it form a parallel class of D. In this way we obtain 7 mutually disjoint parallel classes $\pi_1, \ldots, \pi_7$. Each of the 28 lines of degree 4 is contained in one of these sets. Thus, if we let the lines of π_j intersect in new points ∞_j and if we add the new line $\{\infty_1, \ldots, \infty_7\}$, then we obtain a (7,1)-design with 35 points and 43 lines in which every line has degree 5 or 7. Corollary 13.3 shows that such a design does not exist.■

Now we want to show that $A(5) = 31$. First we need an easy lemma.

<u>13.9 Lemma</u>. Every (6,1)-design with 22 points and 31 lines is extendible.

<u>Proof</u>. We want to show that L has a line of degree $n := 5$. Then Lemma 3.4 shows that L can be extended. Denote by b_k the number of k-lines and assume by way of contradiction that $b_5 = 0$. Since $\sum b_k = b = 31$, $\sum b_k \cdot k = v(n+1)$, and $\sum b_k \cdot k(k-1) = v(v-1)$, it follows that $24b_0+15b_1+8b_2+3b_3 = 18$. In particular $b_1+3b_3 < v$. Hence, there is a point p which is not contained in a line of degree 1 or 3. Every line through p has therefore degree 2, 4, or 6. If t_k denote the number of lines of degree k through p, then it follows that $v-1 = t_2+t_4 3+t_6 5 = 2(t_4+2t_6)+(t_2+t_3+t_4)$, which is not possible, since $t_2+t_4+t_6 = r_p = 6$ and because $v-1$ is an odd number.■

<u>13.10 Theorem</u>. $A(5) = 21$.

<u>Proof</u>. Since the projective plane of order 4 can not be embedded into a projective plane of order 5, $A(5) \geq 21$. Thus, it suffices to show that a non-degenerate linear space L with $22 \leq v \leq b \leq 31$ can be embedded into a projective plane of order $n := 5$. If $b \leq n^2+n$, this follows from Corollary 7.5, and if $b = n^2+n+1$ and some point has degree at most n, it follows from Corollary 12.10. We may therefore assume that every point has degree at least 6 and that $b = 31$. In view of Lemma 13.9 and Corollary 6.23 it suffices to show that every point has degree $n+1$. Assume to the contrary that some point has degree at least 7. We use the notation and the following results of Lemma 7.3.

(1) Every line has degree at most n.

(2) It exists a point of degree $n+1$, and every point of degree $n+1$ is contained in at least 3 lines of degree n.

(3) If a line L has a point of degree $n+1$, then $t_L \leq d_L$.

(4) For every line L there is a line N missing L such that $(d_L n-t_L)k_N \geq (v-k_L)d_L$.

(5) Suppose H and L are two intersecting lines. If a line N of degree n is parallel to H and L, then $d_H d_L \geq n$; in particular, H or L has degree at most $n-2$.

(6) $t_N = 0$ for every n-line, i.e. an n-line has only points of degree $n+1$.

Furthermore,

(7) there is no point of degree $n+2$.

(Assume to the contrary that a point q has degree $n+2 = 7$. Since q is not contained in any line of degree 5 (see (6)), it follows that $v = 22$ and that every line through q has degree 4. Let N be a line of degree n (see (2)). Then q is contained in $r_q - k_N = 2$ lines H and L parallel to N, which contradicts (5).)

Now let N be a line of degree n, and denote by q a point of degree $\neq n+1$. Then $r_q \geq 8$ and q is not contained in any line of degree $n = 5$. Let $\mathcal{M}$ be the set consisting of the 5 lines parallel to N. Then q lies on $\alpha := r_q - 5 \geq 3$ lines of $\mathcal{M}$ which have degree at most 4 (by(6)) and at most one of which has degree 4 (by (5)). It we set $t := \sum (r_p - n - 1)$, then it follows that

$$17 + t \leq v - k_N + t = \sum_{p \notin N} (r_p - n - 1) = \sum_{X \in \mathcal{M}} k_X \leq 4 + (\alpha - 1)3 + (|\mathcal{M}| - \alpha)5 = 26 - 2\alpha.$$

In view of $t \geq r_q - n - 1 \geq 2$ and $\alpha \geq 3$, we obtain $t \leq 3$ and $\alpha = 3$.

We have shown that every point has degree $n+1 = 6$ or $n+3 = 8$. Since every point of degree 8 counts 2 for t, $t \leq 3$ implies that there is a unique point q of degree 8. The point q lies on two lines of degree at most 3 which are parallel to N. In view of $v \geq 22$ we have therefore the following possibilities for the 8 degrees of the lines through q:

3,3,4,4,4,4,4,4 or 2,3,4,4,4,4,4,4 or 3,3,3,4,4,4,4,4.

Thus, q is contained in lines H and L with $k_H \leq k_L = 3$. Let p_1 and p_2 be the two points $\neq q$ of L and let p be a point $\neq q$ of H. Then the lines pp_1 and pp_2 are parallel to at least two of the 4-lines through q and have therefore degree at most 4 (see (5)). Counting the number of points from p, we obtain

$$22 \leq v \leq k_H + 2 \cdot 3 + (r_p - 3) \cdot 4 = k_H + 18.$$

But $k_H \leq 3$, a contradiction. Hence, every point has degree $n+1 = 6$ and L can be embedded into a projective plane of order 5.■

Now we know that $A(2) = 3$, $A(3) = 8$, $A(4) = 15$, $A(5) = 21$, and $A(6) = 31$ and $A(9) = 79$. Since there is a projective plane of order 7, we have $A(8) \geq 7^2+7+1$, so it seems to be possible to determine A(8) (which should be 7^2+7+1). But it seems to be very difficult to determine A(7). Since there is no projective plane of order 6, it is possible that $A(7) < 6^2+6+1 = 43$. We just know $A(7) \geq 31$, because there is a projective plane of order 5.

14. A result on graph theory with an application to linear spaces

In this section we prove a result on graph theory (Lemma 14.2) which will play a crucial role in the next section.

14.1 Definition. A *graph* is a partial linear space in which every line has degree two. The points of a graph will be called vertices. Two distinct vertices which lie on a line are called *joined* or *adjacent*. For every vertex we denote by $V^{\perp}$ the set of vertices which are joined to V (note that $V \notin V^{\perp}$). A *clique* of a graph is a set of pairwise joined vertices.

In other words, a graph consists of a set X of vertices and a relation on X which determines if two points are joined or not.

The following lemma proves the existence of large cliques in a graph under certain conditions. The method of the proof was also used by Bose (1963) to show that certain pseudo-geometric graphs are geometrisable. However the idea of the proof of Lemma 14.2 already appeared in a paper of Bruck (1963).

14.2 Lemma. Let G be a graph and suppose there exists integers a, c, d, e, n, and x with the following properties.

(1) The number of vertices is $dn+x > 0$.

(2) $n-1+a \le |V^{\perp}| \le n-1+c$ for every vertex V.

(3) $|V^{\perp} \cap W^{\perp}| \le e$ for any non-adjacent vertices V,W.

(4) (i) $2n > (d+1)(de-2a)+2x$, or

(ii) There do not exist d+1 pairwise non-adjacent vertices.

(5) $n > (2d-1)c+e-2x$.

(6) $e \ge 0$.

Then we have: If $C_1,\dots,C_s$ are all maximal cliques C with at least $n-(d-1)c+x$ vertices, then $s = d$ and the sets C_i form a partition of the vertex set of the graph.

Proof. In this proof, we call a clique C *normal*, if C is a maximal clique and if $|C| \ge n-(d-1)c+x$. By a *claw* of *order* s we mean a set consisting of s pairwise non-adjacent vertices. A claw of order d is called *normal*. Given any claw S, we denote by $f_S(y)$ the number of vertices outside of S which are adjacent to exactly y vertices of S, $0 \le y \le |S|$. Now we prove the lemma in five steps.

Step 1. If S is claw of order s then $s \le d$ and

$$\sum_{y=0}^{s} f_S(y)y \le s(n-1+c) \quad \text{and} \quad \sum_{y=0}^{s} f_S(y) = nd+x-s.$$

For: Set $f := f_S$ and let T be the set of vertices outside of S. We have

$$\text{(i)} \qquad \sum_{y=0}^{s} f(y) = |T| = dn+x-s.$$

Counting the set $\{(V,W) \mid V \in S,\ W \in T,\ V \text{ and } W \text{ are adjacent}\}$ in two different ways, we obtain

$$\sum_{y=0}^{s} f(y)y = \sum_{V \in S} |V^{\perp}|.$$

Using condition (2), we obtain

$$\text{(ii)} \qquad \sum_{y=0}^{s} f(y)y \le s(n-1+c) \qquad \text{and} \qquad \text{(iii)} \qquad \sum_{y=0}^{s} f(y)y \ge s(n-1+a).$$

It remains to prove $s \le d$. Counting the set $\{(U,V,W) \mid U,V \in S,\ W \in T,\ U \neq V, \text{ and } W$ is joined to U and V$\}$ in two ways, we obtain

$$\sum_{y=0}^{s} f(y)y(y-1) = \sum_{\substack{U,V \in S \\ U \neq V}} |U^{\perp} \cap V^{\perp}|$$

Using condition (3), it follows

$$\sum_{y=0}^{s} f(y)y(y-1) \le s(s-1)e.$$

Combining this inequality with (i) and (iii), we obtain

$$0 \le \sum_{y=0}^{s} f(y)(y-1)(y-2) \le 2(dn+x-s)-2s(n-1+a)+s(s-1)e.$$

If we assume $s = d+1$, then we obtain

$$2n = 2n(s-d) \le 2x-2sa+s(s-1)e = (d+1)(de-2a)+2x.$$

However this contradicts hypothesis (5). Hence there is no claw of order d+1 and therefore also no claw of order at least d+1. This proves Step 1.

Step 2. Every vertex V is contained in a normal claw.

For: Let S be a claw of maximal order s with $V \in S$. Since S is maximal, we have $f_S(0) = 0$. Therefore $\sum f_S(y)y \geq \sum f_S(y)$ and Step 1 shows $(n+c)s \geq nd+x$. Assume by way of contradiction that $s < d$. Since $n+c > 0$ (this follows from hypotheses (1) and (2)), it follows that $(n+c)(d-1) \geq dn+x$ so that $n \leq (d-1)c-x$. Together with hypotheses (5) this yields $dc+e < x$. Hence, $n \leq (d-1)c-x < -c-e$. But $n+c > 0$ and $e \geq 0$, a contradiction.

Step 3. Let S be a normal claw, $V \in S$, and set $S' = S-\{V\}$. Then the set C consisting of all vertices $W \notin S'$ which are not joined to every vertex of S' is contained in a normal clique C'. We have $V \in C'$.

For: Since S' is a claw of order d−1 and because there is no claw of order d+1, the set C is a clique. Since $|C| = f_{S'}(0)$, Step 1 shows that $|C| \geq n+x-(d-1)c$. By definition, any maximal clique C' containing C is therefore a normal clique.

Step 4. Distinct normal clique and are disjoint.

For: Let C_1 and C_2 be distinct normal cliques. Then there exists vertices $V_j \in C_j$ such that V_1 and V_2 are not adjacent. Since any vertex of $C_1 \cap C_2$ is adjacent to V_1 and V_2, hypotheses (3) shows $|C_1 \cap C_2| \leq |V_1^{\perp} \cap V_2^{\perp}| \leq e$.

Assume by way of contradiction that $C_1 \cap C_2$ contains a vertex V. Then $C_1 \cup C_2 - \{V\} \subseteq V^{\perp}$ and hypothesis (2) shows that $|C_1 \cup C_2| \leq n+c$. Consequently $|C_1|+|C_2| = |C_1 \cap C_2| + |C_1 \hat{} C_2| \leq n+c+e$. Because normal cliques have at least $n-(d-1)c+x$ vertices, we conclude $2n-2(d-1)+2x \leq n+c+e$, i.e. $n \leq (2d-1)+e-2x$, which contradicts hypothesis (5).

Now we are in position to prove the lemma. Step 2 and Step 3 show that every vertex is contained in a normal clique, and Step 4 shows that normal cliques are disjoint. It remains to show that there are exactly d normal cliques.

Since there exists vertices, Step 2 shows that there exists a normal claw $S = \{V_1,\ldots,V_d\}$. For $j = \{1,\ldots,d\}$, let C_j be the set of vertices which are not adjacent to any vertex of $S-\{V_j\}$. Step 3 shows that C_j is contained in a normal clique C_j'. Let C be the union of the cliques C_j, $j = 1,\ldots,d$. By definiton, the set C consists of the vertices $V_1,\dots,V_d$ and the vertices which are adjacent to exactly one vertex of S. Hence $|C| = d+f_S(1)$. Since there is no claw of order d+1, we have $f_S(0) = 0$. It follows $f_S(1) \geq 2\sum f_S(y) - \sum f_S(y)y$ (the sums run from $y=1$ to $y=d$), so $f_s(1) \geq nd+2x-d(c+1)$ (use Step 1 with $s = d$). Assume by way of contradiction that there is another normal clique D. Then $D\cap C = \emptyset$ (Step 4). Since nd+x is the number of vertices of the graph and in view of $|D| \geq n-(d-1)c+x$, it follows that

$$dn+x \geq |C|+|D| \geq d+f_S(1)+n-(d-1)c+x \geq n(d+1)-(2d-1)c+3x.$$

Hence $n \leq (2d-1)c-2x$. But this contradicts hypotheses (5) and (6). Therefore $C_1',\dots,C_d'$ are all normal cliques. Since $V_i \in C_i'$ and because V_i and V_j, $i \neq j$, are not joined the cliques C_i' are distinct cliques.■

If V is a vertex of a graph G, then Lemma 14.2 can be applied to the subgraph $G(V^\perp)$ of G which is induced by G on the set $V^\perp$. If the hypothesis of Lemma 14.2 can be satsified for this subgraph, then the conclusion is that there are cliques C_i of G satisfying $V \in C_i$ and $|C_i| \geq n-(d-1)c+x+1$ and that any vertex of $V^\perp$ lies in precisely one of these cliques (notation as in 14.2). We obtain the following application for incidence structures I, if we define G to be the set of lines of I and such that two lines are joined in G if they are disjoint in I.

<u>14.3 Corollary</u>. Let H be a line of an incidence structure I and denote by M the set of lines which are disjoint to H. Suppose there are integers d, e, f, g, n, and x satisfying the following properties.

(1) H is disjoint to $dn+x > 0$ lines.

(2) If L_1 and L_2 are intersecting lines of M, then there are at most e lines which miss L_1, L_2, and H.

(3) If $L \in M$ and if m is the number of lines disjoint to L and H, then $n-1+f \leq m \leq n-1+g$.

(4) (i) $2n > (d+1)(de-2a)+e+2x$, or

(ii) there do not exists d+1 mutually intersecting lines in M.

(5) $n > (2d-1)c+e-2x$

Then there are exactly d sets C of mutually disjoint lines satisfying $H \in C$ and $|C| \geq n-(d-1)c+x+1$. Furthermore every line which is disjoint to H occurs in exactly one of these d sets.■

If the hypotheses of Lemma 14.2 can be satisfied for every line of a linear space, then hypothesis (2) of the following embedding theorem is satisfied. We remark that this theorem has been proved in (Beutelspacher and Metsch, 1986) for linear spaces with maximal point degree $n+1$, however the proof was much more complicated than the one presented here.

14.4 Theorem. Suppose L is a partial linear space and there exists an integer $n \geq 2$ such that the following three conditions hold.

(1) $b \geq n^2$.

(2) For every line H there exists an integer $t(H)$ with the following properties: It exists exactly $n+1-k_H$ sets M of pairwise disjoint lines with $H \in M$ and $|M| \geq t(H)$. Furthermore, every line which is disjoint to H appears in exactly one of these sets M.

(3) It exists a point of degree $n+1$ which is joined to every other point.

Then I can be extended to a projective plane of order n.

Remark. The condition $b \geq n^2$ can be replaced by the condition $b \geq N(n)$, where $N(n)$ denotes the least integer such that every partial projective plane with at least $N(n)$ lines can be extended to a projective plane of order n. This follows easily from the following proof.

Proof of Theorem 14.4. We call a maximal set M of mutually disjoint lines a *clique* if it satisfies $|M| \geq t(L)$ for at least one line L of M. Let X be the set of all cliques. First we want to show the following.

Claim. If M is a clique, then $|M| \geq t(L)$ for every line $L \in M$.

Proof. Suppose it exists a clique M and a line L with $L \in M$, $|M| < t(L)$. By definition, M contains a line L' with $|M'| \geq t(L')$. Then L and L' are parallel. By hypothesis (2), it exists a clique M' with $L, L' \in M'$ and $|M'| \geq t(L)$. But now M and M' are distinct cliques which contain L and L' and have at least $t(L')$ lines. This contradicts hypothesis (2) used for L'.

An immediate consequence of the claim and hypothesis (2) is that every line L lies in exactly $n+1-k_L$ cliques and that disjoint lines are contained in exactly one clique. Now we extend the incidence structure L in the following way. For every clique M we adjoin a new point p_M which should be incident with the lines of M. In this way we obtain an incidence structure L' which has the property that any two of its lines intersect in a point. If we can show that L' has n^2+n+1 points, then L' is therefore a partial projective plane. Since L' has at least n^2 lines, it follows then from (the result dual to) Lemma 3.4 that L' (and hence also L) can be extended to a projective plane of order n.

It remains to show that L' has n^2+n+1 points. By hypothesis (3) there exists a point p of degree n+1. Let $L_0,\ldots,L_n$ be the lines through p, denote the degree of L_i by $n+1-d_i$, and set $D := d_0+\ldots d_n$. Since p is joined to every point, we have $v = 1+r_pn-D = n^2+n+1-D$. The line L_i lies in d_i cliques. Hence there are D cliques containing one of the lines L_i. In order to show that L' has n^2+n+1 points, we have therefore to show that every clique contains a line through p.

Suppose L is a line with $p \notin L$. Then p lies on at least $d := n+1-k_L$ lines which miss L. Any of these lines together with L determines a clique. Since L lies in d cliques, it follows that every clique which contains L contains also a line through p. Since this holds for every line L (with $p \notin L$), it follows that every clique contains a line through p.■

The preceding theorem together with Lemma 14.3 was used in by Beutelspacher and Metsch (1986, 1987) to obtain the following results.

<u>14.5 Result</u>. Every linear space with maximal point degree n+1 and minimal line degree n+1−a can be embedded into a projective plane of order n provided that $4n > 6a^4+9a^3+19a^2+8a$.

<u>14.6 Result</u>. Suppose L is a linear space with constant point degree $n+1 \geq 3$, maximal line degree n+1−a, and minimal line degree n+1−A.

1) If $b \leq n^2+n+1$ and $2n > (A^2-1)(A^2+A-2a+2)$, then L can be embedded into a projective plane of order n.

2) If $b > n^2+n+1$, then $2n \leq (A^2-1)(A^2+A-2a+2) + A(A-1)(A-a-1)a$.

We prove another lemma, which will be needed in the next section.

14.7 Lemma. Let u, s, and n be non-negative integers satisfying $n > 2u+s-2$ and $n > u+s^2-1$, and suppose it exists a projective plane P of order n and a set M of points of P satisfying the following properties.

(1) $|M| = u(n+1)+s$.

(2) Every line of P intersects M in at least u points.

Then $u \in \{0,1\}$ and M contains exactly u lines.

Proof. If L is any line which has a point p outside of M, then

$$un+u+s = |M| = \sum_{p\in X} |X\cap M| \geq |L\cap M|+(r_p-1)u = |L\cap M|+un.$$

It follows that $|L\cap M| \leq u+s$. Hence $L \subseteq M$ or $u \leq |L\cap M| \leq u+s$ for every line L. Now let q be a point of M and denote by z the number of lines L through q with $L \subseteq M$. Then

$$un+u+s-1 = |M-\{q\}| = \sum_{q\in X} (|X\cap M|-1) \geq zn+(n+1-z)(u-1),$$

which implies that $z(n+1-u) \leq n+s$. In view of $n \geq 2u+s-1$, it follows that $z \leq 1$. Hence, every point of M lies on at most one line L with $L \subseteq M$. This implies that there is at most one line contained in M.

For every line L set $x_L = |L\cap M|-u$. Then count pairs (q,L) with points q of M and lines L satisfying $q \in L$, and count triples (q,q',L) with distinct points q,q' of M and lines L satisfying $q,q' \in L$ to obtain

$$\sum_L (u+x_L) = |M|(n+1) = (un+u+s)(n+1) \quad \text{and}$$

$$\sum_L (u+x_L)(u+x_L-1) = |M|(|M|-1) = (un+u+s)(un+u+s-1)$$

which implies that

$$\sum x_L = un+(n+1)s \quad \text{and} \quad \sum x_L^2 = un^2+(u+s-u^2)n+s^2.$$

We know that $0 \leq x_L \leq s$ or $x_L = n+1-u$ for every line L.

First assume that M contains no line. Then $0 \leq x_L \leq s$ for every line L. It follows that

$$un^2+(u+s-u^2)n+s^2 = \sum x_L^2 \le \sum x_L \cdot s = uns+(n+1)s^2,$$

which implies that $un+u+s-u^2 \le us+s^2$. Using $n > u+s^2-1$, we conclude that $(u-1)(s^2-s) < 0$. It follows that $u = 0$, which we had to show.

Now suppose that there exists a (unique) line $H \subseteq M$. We have to show that $u = 1$. In view of $H \subseteq M$, we have $n+1 \le |M| \le u(n+1)+s$, which implies that $u \neq 0$, since $n > u+s^2-1$. Since $x_H = n+1-u$ and $x_L \le s$ for every other line L, we have

$$un^2+(u+s-u^2)n+s^2 = \sum x_L^2 \le (n+1-u)^2+[(\sum x_L)-(n+1-u)]s$$
$$= (n+1-u)^2+[un+s(n+1)-n-1+u]s,$$

which implies that $(u-1)n^2 \le (u-1)(un-2n+u+1+ns+s)+ns(s-1)$. Assume by way of contradiction that $u \ge 2$. Then it follows that

$$n^2 \le un-2n+u+1+ns+s+ns(s-1) = n(u-2+s^2) + u+s+1.$$

In view of $u+s+1 \le 2u+s-1 \le n$, this implies that $n \le s^2+u-1$, a contradiction. Hence, $u = 1$ and the lemma is proved.■

An easy application of the preceding lemma (and its proof), is the following result due to A. Bruen (1971). Recall that a blocking set of a projective plane is a set B of its points such that every line meets B but no line is contained in B.

<u>14.8 Result</u>. Suppose B is a blocking set of a projective plane of order n. Then $|B| \ge n+\sqrt{n}+1$ with equality if and only if B is the set of points of a Baer-subplane of P (which implies that n is a perfect square).

<u>Proof</u>. Denote the number of points of B by $n+1+s$ and suppose that $s^2 \le n$. We have to show that equality holds and that B is the set of points of a Baer-subplane.

Set $u = 1$ so that B has $u(n+1)+s$ points. Since B contains no line, Lemma 14.6 shows that $n \le u+s^2-1 = s^2$ or $n \le 2u+s-2 = s$. Consequently, $n = s^2$. As in the proof of the preceding lemma, we set $|L \cap B| = 1+x_L$ for every line L and we obtain the inequality $un^2+(u+s-u^2)n+s^2 = \sum x_L^2 \le \sum x_L \cdot s = uns+(n+1)s^2$. In view of $u = 1$ and $n = s^2$, equality holds, which implies that $x_L^2 = x_L \cdot s$ for every line L. Hence $x_L \in \{0,s\}$ for every line L, i.e. every line meets B in 1 or $1+\sqrt{n}$ points. It is easy to see now that B must be the set of points of a Baer-subplane.

Remark. Since a Baer-subplane of a projective plane P of order n has u(n+1)+s points (with u = 1 and $n = s^2$) and because a Baer-subplane contains no line of P, we see that the bound $n > u+s^2-1$ in Lemma 14.7 is best possible in general. Also the bound $n > 2u+s-2$ is best possible as the following example shows.
Let P be a projective plane of even order n and suppose L is a set of n+2 lines such that every point lies in exactly two or no line of L (such a set exists in the desarguesian projective planes of order 2^t). Let M be the set of points which lie on two lines of L, and set s = 0 and u = (n+2)/2. Then $|M| = u(n+1)+s$ and $n = 2u+s-2$. Furthermore every line which is not in L meets M in exactly u points but there are $|L| = n+2 = 2u$ lines which are contained in M.

Remark. The graph theoretical results and its applications to linear spaces have recently been improved in (Metsch, 1991d). For example, it was shown that it suffices to demand $n > ca^3$ for some constant c in Result 14.5. However, the essential result is that there is a polynomial function f(a) such that $n > f(a)$. This is one reason that the results of this section have not been updated. Also, the earlier results will suffice for the application in the next section.

15. Linear spaces in which every long line meets only few lines.

An interesting property of linear spaces was discovered by Hanani (1954/55) (see also Varga, 1985): Every line L of maximal degree of a linear space meets at least v lines. This generalizes the Theorem of De Bruijn and Erdös (Theorem 2.3). Motivated by this property, we denote by c(L) the least integer such that every line L of maximal degree of a linear space L meets at most v+c(L) lines (notice that each line meets itself). In other words, v+c(L) is the maximal number of lines which meet a line of maximal degree. Then $c(L) \geq 0$ for every linear space L. De Witte (1975b) proposed to study linear spaces L with small values c(L). This was done by Melone (1987) for $c(L) = 1$ and De Vito/Lo Re (1988a,b) for $c(L) = 2$. It was conjectured (De Vito/Lo Re/Metsch, 1991) that for any positive integer c all except a finite number of linear spaces with $c(L) = c > 0$ can be obtained from a projective plane by removing c points or from an affine plane by removing c−1 points. This conjecture has been proved in (Metsch, 1990b). Here we present this result only in an asymptotic form, which has the advantage that tedious computations in the proof can be avoided.

We begin studying linear spaces in which the minimal point degree is less than the maximal line degree. This case provides a nice example for an application of the Lemma of Stanton and Kalbfleisch (Lemma 2.4).

15.1 Lemma. Suppose that L is a linear space with v points, minimal point degree r and maximal line degree k satisfying $r < k$. Let L be a line of maximal degree and denote by v+c the number of lines which meet L. If $k > c+2$, then L is a near-pencil.

Proof. The hypotheses imply that $k \geq c+2$. Hence $v \leq 1+r(k-1) \leq 1+(k-1)^2 = k^2-k-c-(k-2-c) \leq k^2-k-c$. The Lemma of Stanton and Kalbfleisch shows that L meets at least $k^2(v-k)/(v-1)$ other lines. It follows that $(v-1+c)(v-1) \geq k^2(v-k)$, which can be transformed into $1+k^3-c \geq v(k^2+2-c-v) =: g(v)$.

In order to prove the lemma, we have to show that $v = k+1$. Assume to the contrary that $k+2 \leq v \leq k^2-k-c$. Since g is a polynomial of degree two with negative coefficient in v^2, we have $g(v) \geq g(k+2) = g(k^2-k-c) = (k+2)(k^2-k-c)$. Hence $1+k^3-c \geq (k+2)(k^2-k-c)$, which implies that $(k+1)c \geq (k+1)(k-3)+2$. It follows that $k < c+3$, a contradiction.■

As an application of Lemma 15.1 we prove the result mentioned above.

15.2 Theorem (Hanani 1954/55, Varga 1985). Let L be a linear space, v its number of points, and let L be a line of maximal degree. Then L meets at least v−1 other lines with equality if and only if L is a generalized projective plane.

Proof. If L is a generalized projective plane then every line of maximal degree meets obviously v−1 other lines.

Suppose now that L is a line of maximal degree which meets at most v−1 other lines. We denote the minimal point degree by r and the maximal line degree by k.

If $k > r \geq 2$, the Lemma 15.1 shows that L is a near-pencil (since we have $c \leq 0$ in 15.1 so that $k > c+2$). Assume from now on that $r \geq k$.

The line L of maximal degree k meets at least k(r−1) other lines with equality only if every point of L has degree r. Since L by hypothesis meets at most v−1 other lines, it follows that $k(r-1) \leq v-1$ with equality only if each point of L has degree r. On the other hand, we have $v-1 \leq r(k-1)$ with equality only if an r-point lies only on k-lines. It follows that $k(r-1) \leq v-1 \leq r(k-1)$, i.e. $r \leq k$. Hence $r = k$ and equality holds in the above inequalities. Therefore every point of L has degree r and every line through a point of L has degree $k = r$. We have shown that there are v lines which meet L (L is one of these lines!) and that these lines have all degree k. Since $vk(k-1) = v(v-1)$, it follows that these v lines already cover all pairs of points of L. This implies that there is no other line. Consequently, every line meets L and has degree k. This implies that every point and line has degree $k = r$. Thus L is a projective plane of order k−1 (or the near-pencil on three points if $k = 2$).■

15.3 Theorem. For every sufficiently large value for c we have the following. Every non-degenerate linear space with maximal line degree $k \geq 7c^7$ in which every line of maximal degree meets at most v+c lines can be obtained from an affine plane by removing at most c−1 points or from a projective plane by removing at most c points.

Before we start with the proof of this theorem, we prove an easy corollary.

<u>15.4 Corollary</u>. For every sufficiently large value for c we have the following. Every non-degenerate linear space with $v \geq (7c^7+c)^2$ points in which every line of maximal degree meets at most v+c lines can be obtained from an affine plane by removing at most c−1 points or from a projective plane by removing at most c points.

<u>Proof</u>. Let r be the minimal point degree and k the maximal line degree. Then a k-line meets at least k(r−1) other lines and the number of points is $v \leq 1+r(k-1)$. Since a k-line meets at most v−1+c other lines, it follows that $k(r-1) \leq v-1+c \leq r(k-1)+c$, i.e. $r \leq k+c$. Hence $v-1 \leq r(k-1) \leq (k+c)(k-1)$. The hypothesis on v implies therefore that $k \geq 7c^7$, and the corollary follows from Theorem 15.3.■

<u>Remarks</u>. 1. It follows from Theorem 15.4 that for every value of c there are only a finite number of non-degenerate linear spaces in which v+c is the maximal number of lines meeting a line of maximal degree and which is not affine plane with c−1 points deleted or a projective plane with c points deleted.

2. It was shown in (Metsch, 1990b) that it suffices to require $k > 7c^7+40c^6-c$ in Theorem 15.3 for every $c \geq 2$.

In the rest of this section we shall prove Theorem 15.3. Throughout we assume that L is a non-degenerate linear space with maximal line degree k in which every k-line meets at most v+c lines. We assume that $k \geq 7c^7$ and $c \geq 2$ (later on, we shall restrict ourselfes to larger values for c). We denote the minimal point degree by r. A point will be called *real* if it has degree r, and a *real* line is a line with a real point. Furthermore, we shall use the following notation.

s is defined by $v = 1+r(k-1)-s$,

b_r denotes the number of real lines,

$n := r-1$, $z := b_r-(n^2+n+1)$, $w := 3c(c+1)/2$,

for every line L we set $d_L = n+1-k_L$ and $t_L = \sum_{p \in L} (r_p-r)$,

$t := \max \{ t_L \mid L \text{ is a k-line} \}$, and

$G := \{L \mid L \text{ is a real line with at most } 1+r/w \text{ non-real points}\}$.

Finally, I denotes the incidence structure consisting of the real lines of G and of all points of L.

Using Corollary 14.3 and the embedding theorem 14.4 we shall show that I can be extended to a projective plane P. We shall denote by M the set of points of P−I and show that M satisfies the assumptions of Lemma 14.6 so that M consists of a few number of points and (eventually) one line. Finally, we shall show that L is the complement of M in P, which will complete the proof of the theorem.

15.5 Lemma. $t+s+(r-k) \le c$, $r \ge k$, $s \ge 0$ and $t \ge 0$.

Proof. Since L is non-degenerate, we have $r \ge k$ (see 15.1). Since r is the minimal point degree and k is the maximal line degree, we have $v \le 1+r(k-1)$ so that $s \ge 0$. Since r is the minimal point degree, the definition of t implies that $t \ge 0$.

By hypothesis, a line L of maximal degree k meets at most $v+c = 1+r(k-1)-s+c$ lines. By definition of t_L, the line L meets $1+k(r-1)+t_L$ lines. Thus $1+r(k-1)-s+c \ge 1+k_L(r-1)+t_L$ so that $t_L+s+r-k \le c$. Since $t = \max\{ t_L \mid L \text{ a } k\text{-line}\}$, it follows that $t+s+r-k \le c$.■

15.6 Lemma. Every point of degree r lies on at least $r-s \ge r-c$ lines of degree k, and every line of degree k has at least $k-t \ge r-c$ points of degree r. Furthermore, every real line has degree at least $k-s \ge r-c$.

Proof. Let p be an r-point and denote by α the number of lines of degree k through p. Then $v-1 \le \alpha(k-1)+(r_p-\alpha)(k-2) = r(k-2)+\alpha$. Since $v-1 = r(k-1)-s$, it follows that $\alpha \ge r-s$. By the definition of t, every k-line has at most $t_L \le t$ points which have not degree n+1.

If H is a real line and q a real point of H, then we have $v \le k_H+(r_q-1)(k-1)$, which implies that $k_H \ge k-s$.■

15.7 Lemma. Let H be a real line and denote by nd_H+y the number of real lines disjoint from H. Then $y = O(c^2)$ (this means that there is a constant M which does not depend on c or n such that $|y| \le Mc^2$).

Proof. By definition, H has a real point. Lemma 15.6 shows therefore that H meets a k-line $L \ne H$. If p is a real point $\ne L\cap H$ of L, then p lies on $d := d_H$ lines which miss H and these lines are real. Since L has at least k−t real points, it follows that H is disjoint from at least $(k-1-t)d = (r-1)d-(r-k+t)d$ real lines. Lemma 15.5 implies $y \geq -(r-k+t)d \geq -cd \geq -c^2$.

Let q be a real point of H and let w be the number of k-lines $\neq H$ through q. We shall find a lower and an upper bound for the number x of pairs (K,R) consisting of a k-line $K \neq H$ through q and a real line R which meets K and misses H. If K is a k-line, then each point $p \in K-\{q\}$ lies on $r_p-k_H = d+r_p-r$ lines missing H. Hence K meets $(|K|-1)d+t_K \leq (r-1)d+t_K = nd+t_K \leq nd+t$ lines which miss H. Consequently $x \leq (r_q-1)(nd+t) = n(nd+t) \leq n(nd+c)$.

Now let H be one of the nd+y real lines which miss H. Since q lies on at most c lines of degree $\neq k$, the line L meets at least $k_L-c \geq r-2c = n+1-2c$ (Lemma 15.6) of the k-lines K through q. Hence $(nd+y)(k-2c) \leq x$.

Together we obtain $(nd+y)(n+1-2c) \leq n(nd+c)$. Since $n+1 = r \geq 7c^7$ and $d \leq c$, it follows that $y \leq (3c+1)d \leq (3c+1)c$, proving the lemma.■

We recall that the number of real lines is denoted by $b_r = n^2+n+1+z$.

<u>15.8 Lemma</u>. We have $z = O(c^2)$.

<u>Proof</u>. Let L be a k-line and set $d := r-k$. We have $n^2+n = k(r-1)+dn$ (recall that $n = r-1$). Since the number of real lines missing L is $nd+O(c^2)$ by 15.7, we have to show that the number of real lines meeting L is $k(r-1)+O(c^2)$. Since L meets $k(r-1)+t_L$ other lines, it is trivial that L meets at most $k(r-1)+O(c^2)$ real lines.

Let q be a real point of L and let H be a second k-line through q. Lemma 15.6 says that L and H have at most c non-real points. Each real point of L lies only on real lines, and each non-real point of L is joined to the real points of H by a real line so that it lies on at least k-c real lines. If e is the number of non-real points of L, then it follows that L meets at least $(k-e)(r-1)+e(k-1-c) = k(r-1)-e(r-k+c)$ other real lines. Since $e \leq c$ and $r-k \leq c$ (see 15.5), this proves the lemma.■

<u>15.9 Lemma</u>. Every point lies on at most r+s real lines.

<u>Proof</u>. Let m be the number of real lines passing through a point p. Since every real line has at least k-s points, we have $m(k-1-s) \leq v-1 = r(k-1)-s$.

Assume that $m \geq r+1+s$. Then it follows that $k \leq 1+s(s+1+r-k) \leq 1+c(c+1)$ (see Lemma 15.5), a contradiction.■

15.10 Lemma. If K and L are intersecting real lines, then there the number of real lines which miss K and L is $O(c^2)$.

Proof. By 15.7 and 15.8, the line K meets $b_r-d_K n+O(c^2) = n^2+n-d_K n+O(c^2)$ real lines. Similarly, the line L meets $n^2+n-nd_L+O(c^2)$ real lines. Let α be the number of real lines through $q := K \cap L$. Then there are at most $(k_K-1)(k_L-1)+\alpha = (n-d_K)(n-d_L)+\alpha = n^2-(d_K+d_L)n+\alpha+O(c^2)$ which meet K and L. Hence there are at least $n^2+2n-\alpha+O(c^2)$ real lines which meet K or L so that 15.8 implies that there are at most $\alpha-n+O(c^2)$ lines which miss K and L. Since $\alpha \leq r+s = n+1+s$ (see 15.9), this proves the lemma.■

15.11 Lemma. Suppose H and L are parallel real lines and $n-1+y$ is the number of real lines parallel to H and L. Then $y \leq O(c^2)$ (hereby we mean that there is a constant M not depending on c or n such that $y \leq Mc^2$). If H and L have at most r/w non-real points, then $y \geq O(c^2)-rc/w$ (we recall that $w := 3c(c+1)/2$).

Proof. As in the previous lemma, there are $n^2+n-d_H n+O(c^2)$ real lines which meet H and $n^2+n-nd_L+O(c^2)$ real lines which meet L. Since there are at most $k_H k_L = (n+1-d_H)(n+1-d_L) = n^2+2n-(d_H+d_L)n+O(c^2)$ real lines which meet H and L, it follows that there are at least $n^2+O(c^2)$ lines which meet H or L. Now 15.8 shows that the number of real lines which miss H and L is at most $n+O(c^2)$. This proves the first part.

Now we suppose that H has μ_H non-real points and L has μ_L non-real points with $\mu_H \leq 1+r/w$ and $\mu_L \leq 1+r/w$. Let S be the set of points of L which do not lie on a k-line. By Lemma 15.6, the points of S are non-real. It follows from 15.6 that there exists a real point p_0 outside of L and that p_0 lies on at least r-s lines of degree k. Hence $|S| \leq s$.

Let p be a point of L. We want to find an upper for the number x of lines $\neq$ L through p which miss H. If p is a real point then $x \leq (r_p-1)-k_H = d_H-1$. If p is not real but lies on a k-line G then $r_p \leq r+t_G \leq r+t$ and $x \leq (r_p-1)-k_H = d_H-1+t$. If $p \in S$, then p lies on at most r+c real lines (Lemma 15.9) and p is joined to the $k_H-\mu_H$ real points of H by a real line. Hence we have $x \leq r+c-1-(k_H-\mu_H)$

$= d_H+c-1+\mu_H$ in this case. It follows that the number of real lines $\neq L$ which meet L and miss H is at most

$$(k_L-\mu_L)(d_H-1) + (\mu_L-|S|)(d_H-1+t) + |S|(d_H-1+c+\mu_H)$$
$$= k_L(d_H-1) + \mu_L t + |S|\mu_H + |S|(c-t)$$
$$\leq k_L(d_H-1) + \mu_L t + s\mu_H + sc$$

Since $\mu_L t+\mu_H s \leq (t+s)(1+r/w) \leq c(1+r/w)$, $k_L \leq n+1$ and $d_H \leq c$, it follows that there are at most $n(d_H-1)+cr/w+O(c^2)$ which meet L and miss H. Since H misses $d_H n+O(c^2)$ real lines, it follows that the number of real lines which miss L and H is at least $n-rc/w+O(c^2)$.■

<u>15.12 Lemma</u>. There are at most c^2w^2 real lines with more than $1+r/w$ non-real points.

<u>Proof</u>. Let p be a real point. By Lemma 15.6, the point p lies on (at least) r−s lines $L_1,\ldots,L_{r-s}$ lines of degree k. By definition of t, the lines L_j have at most t non-real points. Hence, if Q is the set of non-real points lying on one of the lines L_j, then $|Q| \leq (r-s)t$. Let M be the set of lines L with $p \notin L$ and with at least $1+r/w$ non-real points. Then every line of M has at least $1+r/w-s$ points in Q. Counting triples (q,q',L) with distinct points $q,q' \in Q$ and lines $L \in M$ satisfying $q,q' \in L$ shows that

$$|M|(\tfrac{r}{w}-s)^2 \leq \sum_{L\in M} |L\cap Q|(|L\cap Q|-1) \leq |Q|(|Q|-1) \leq (r-s)^2t^2.$$

Consequently $|M|(r-ws)^2 \leq (r-s)^2w^2t^2 = [(r-ws)^2+s(2rw-w^2s-2r+s)]w^2t^2$, which implies that $(|M|-w^2t^2)(r-ws)^2 \leq 2srw^3t^2$. Since $r-ws \geq r-wc \geq r/2$, we conclude that $|M|-w^2t^2 \leq 8sw^3t^2/r$. Since $w = 3c(c+1)/2$ and $t \leq c$, we have $8wt^2 \leq 7c^7 \leq r$ (since $c \geq 2$). It follows that $|M| \leq w^2t^2+sw^2$. Since $t+s \leq c$ and $c \geq 2$, we have $t^2+s \leq (c-s)^2+s = c^2-2cs+s^2+s \leq c^2-s$ and hence $|M| \leq w^2(c^2-s) \leq w^2c^2-s$. Every k-line through p has at most $t \leq c$ non-real points and hence p lies on at most s lines with at least $1+r/w$ non-real points. This proves the lemma.■

Now we are almost in position to prove that I can be extended to a projective plane. Recall that I is the incidence structure consisting of all points of L and of the real lines of L which have at most $1+r/w$ non-real points.

<u>15.13 Lemma</u>. The incidence structure I has at least $n^2+n+1-n/3c$ lines and at least one point of degree $n+1$.

<u>Proof</u>. L has $n^2+n+1+O(c^2)$ real lines (Lemma 15.8), so I has at least $b_r-w^2c^2+O(c^2)$ real lines (Lemma 15.12). Since $r = n+1 \geq 7c^7$ and $w = 3c(c+1)/2$, we see that for sufficiently large c the number of lines of I is at least $n^2+n+1-n/3c$.

Let L be a k-line of L. Then L has at least $k-t \geq r-c$ real points. Since there are at most $w^2t^2 < r-c$ real lines which are not lines of I (Lemma 15.12), it follows that L has a real point which lies only on real lines belonging to I. This point has degree $r = n+1$ in I.■

<u>15.14 Lemma</u>. The incidence structure I can be extended to a projective plane P of order n.

<u>Proof</u>. This will follow from the preceding lemma and the embedding theorem 14.4, if we can show that for each line H of I there exists an integer t(H) with the following property: There are exactly d_H sets M of mutually disjoint lines of I with $|M| \geq t(H)$ and $H \in M$; furthermore every line of I which is disjoint from H is contained in exactly one of these sets M.

Let H be a line of I. The existence of t(H) will follow from Lemma 14.3. In order to apply this lemma, we denote by D the set of lines of I which are disjoint from H and we set $d := d_H$ and $x := |D|-dn$. In view of 15.7, 15.10 and 15.10, there exists a constant M (independent of c) such that the integers

$$e := Mc^2 \text{ and } f := -(Mc^2+rc/w+w^2c^2).$$

satisfiy the following conditions

1) $-w^2c^2-Mc^2 \leq x \leq Mc^2$.

2) If K,L are intersecting lines of D, then at most e lines of D miss K and L.

3) If $L \in D$ and if α is the number of lines in D which miss L, then $n-1+f \leq \alpha$ and $\alpha \leq n-1+e$.

In order to apply Lemma 14.3, it remains to show that (notice that we have replaced a resp. c of 14.3 by f resp. e)

4) $2n > (d+1)(de-2f)+2x$, and

5) $n > 2de-2x$.

Since $e \geq 0$, $f \leq 0$, $g \geq 0$, and $d = d_B \leq c$, it suffices to verify these inequalities in the case $d = c$. Then 5) holds, since $n+1 = r \geq 7c^7 \geq 2ce+2c^2w^2$ for sufficiently large values for c. Ignoring terms of magnitude at most c^4, condition 4) requires that $2r = 2(n+1) > 2(c+1)(w^2c^2+rc/w)$ or equivalently

$$r\left(1-\frac{c(c+1)}{w}\right) > (c+1)w^2c^2.$$

Since $w = 3c(c+1)/2$, this is equivalent to $4r > 27c^4(c+1)^3$. Since $r \geq 7c^7$, it follows that condition 4) is satisfied for sufficiently large values for c. This completes the proof of the Theorem 15.3.■

Lemma 15.14 says that I can be embedded into a projective plane P of order n. From now on, we denote by M the set of points of P which are not points of I. Since I has $v = 1+r(k-1)-s$ points, we have $|M| = (n+1)u+s$ where $u := r-k$.

15.15 Lemma. Every point p of L has degree at most n+1+c in L and it has degree at least n+1-n/3c in I.

Proof. Let p be a point of L. Then p is also a point of I. Since I is embedded in P and because I has at least $n^2+n+1-n/3c$ lines (Lemma 15.12), the point p has degree at least $n+1-n/3c$ in I (and therefore also in L). If p lies in L on a k-line L, then p has in L degree at most $n+1+t_L \leq n+1+t \leq n+1+c$. It suffices therefore to show that p lies on a k-line.

Assume to the contrary that p does not lie on a k-line, and consider p as a point of I. Since every line of I through p has in L degree at most $k-1 = n-u$, it follows that every line of I through p meets M in at least u+1 points. Since p has degree at least $n+1-n/3c$ in I, it follows $(n+1-n/3c)(u+1) \leq |M| = (n+1)u+s$. This implies that $3c(n+1-s) \leq n(u+1) \leq n(c+1)$, which is a contradiction.■

15.16 Lemma. Suppose L_0 is a line of P with $|L_0 \cap M| \leq u+s$. Then the set $X := L_0-M$, which consists of the points of L_0 which are in the linear space L, is actually the set of points of a line of L.

Proof. Assume to the contrary that X is not the set of points of a line of L. We consider first the case that the space L has a line L which contains all points of X and a point p which is not in X. Then the point p is a point of P which is not on L_0. For $x \in X$, the line px of the space **L** is the line L. But in **P**, the lines px for $x \in X$ are all different, since $X \subseteq L_0$ and $p \notin L_0$. It follows that the lines px, $x \in X$, of **P** are not lines of **I**. Hence p has degree at most $n+1-|X| \leq u+s \leq c$ (see Lemma 15.5) in **I**. Lemma 15.15 shows that this is not possible.

Now we consider the case that **L** has no line L with $X \subseteq L$. This implies that L_0 is not a line of **I** or **L**. Let $\mathcal{B}$ be the set of lines of **L** which have at least two points in X. Then $|\mathcal{B}| \geq 2$. It follows that the lines of $\mathcal{B}$ induce a linear space $L(X) = (X,\mathcal{B})$ on the points of the set X. The line L_0 meets M in at most u+s points. Since **I** has at least $n^2+n+1-n/3c$ lines, it follows that there are two points p_1 and p_2 in $X = L_0-M$ which have degree n in **I** (i.e. every line $\neq L_0$ of **P** through p_j is a line of **I**). If H is a line of **I** through p_j, then $|H \cap X| = |H \cap L_0| = 1$ and therefore H is not a line of the L(X). Since p_j has degree at most $n+1+c$ (Lemma 15.15), it follows that p_j has degree at most $c+1$ in L(X). Hence, every line $\neq p_1p_2$ of L(X) through p_1 has degree at most $c+1$ in L(X). Counting the number $|X|$ of points of L(X) using the lines through p_1 shows therefore that $|X| \leq y+c^2$ where y is the degree of the line p_1p_2 in L(X). It follows that $y \geq |X|-c^2 = n+1-|L_0 \cap M|-c^2 \geq n+1-c-c^2$. Let p be a point of X which is not on the line p_1p_2 of L(X). Then p has degree at least y in L(X). Since p has degree at least $n+1-n/3c$ in **I** and because every line of **I** through p meets X only in p (use the same argument we used for the points p_j), it follows that p has degree at least $y+n+1-n/3c$ in **L**. But every point of **L** has degree at most $n+1+c$, so $y+n+1-n/3c \leq n+1-c$, which contradicts $y \geq n+1-c-c^2$.■

Now we can easily complete the proof of Theorem 15.3. Suppose L is a line of **P** with $|L \cap M| \leq u+s$. Then the preceding lemma shows that L−M is the set of points of a line of **L**. Since every line of **L** has degree at most $k = n+1-u$, it follows that $u \leq |L \cap M|$. Hence every line of **P** meets M in at least u points. In view of $|M| = (n+1)u+s$, Lemma 14.7 shows now that $u = 0$, or that $u = 1$ and that M contains a line. In both cases it follows that every line L of **P** is either contained in M or satisfies $|L \cap M| \leq u+s$.

If L is a line of **P** which is not contained in M, then the preceding lemma shows that L−M is the set of points of a line of **L**. Hence, the complement P−M of M in **P** is a subspace of the linear space **L**. Since **L** and P−M have the same set of points, we have actually $L = P-M$. Hence, if $u = 0$, then L is the complement of s points in P. If $u > 0$, then $u = 1$ and M contains a line L, and L is the complement of s points in the affine plane P−L. Since $s \le c-u$ (Lemma 15.5), this completes the proof of Theorem 15.3.■

<u>Problem</u>. What is the smallest real number α such that there is an absolute constant t satisfying the following condition: If L is any non-degenerate linear space L, if $c := c(L)$, and if k is the maximal line degree, then $k \ge tc^{\alpha}$ implies that L can be obtained from a projective plane by removing c points, or from an affine plane by removing c−1 points. Theorem 15.3 shows that $\alpha \le 7$. However it seems very likely that this bound is far from being the best possible. A lower bound can be obtained from the complement of a Baer-subplane in a projective plane. In this case we have $k = (c-1)^2$ so that $\alpha \ge 2$.

16. s-fold inflated projective planes

The last aim in this paper is the classification of all linear spaces satisfying $b \leq v+r_q-2$ for some point q, and the proof of the Dowling-Wilson Conjecture. This will be done in section 17. One of the main tools we use is a characterization of s-fold inflated projective planes, which we shall obtain in this section.

Let $L = (p,L)$ be a linear space. A *subspace* of L is a linear space $L_1 = (p_1,L_1)$ with $p_1 \subseteq p$ and $L_1 \subseteq L$. Suppose that L has $s \geq 1$ subspaces $D_j = (p_j,L_j)$, $j = 1,\dots,s$, and that it exists a point q with $p_i \cap p_j = \{q\}$ for $i \neq j$. If $s = 1$ then we suppose furthermore that $q \in p_1$ and $D_1 \neq L$. Then we can smooth L in the way that we replace D_j by a line (i.e. we remove the lines of D_j and adjoin the set p_j as a new line), $j = 1,\dots,s$, and obtain a linear space L'. Suppose that L' has n^2+n+1 lines and that L' can be extended to a projective plane P of order n. Then we call L an *s-fold inflated projective plane* of *order* n. In L' the point q lies on the lines p_j and on n+1-s further lines $L_0,\dots,L_{n-s}$. If $n+1-d_j$ is the degree of L_j in L then $d_0+\dots+d_{n-s}$ is called the *deficiency* of L. The spaces D_j are called the *main subspaces* of L. Hence, an s-fold inflated projective plane with deficiency d can be obtained from a projective plane of order n as follows: We fix a point q and s lines $H_1,\dots,H_s$ through q. For $j \in \{1,\dots,s\}$, we remove the line H_j and some of the points of H_j (but not q) and impose a linear space D_j on the remaining points of H_j. Furthermore, we remove d points which do not lie on any line H_j in such a way that we do not produce lines of size less than 2. Notice that a 1-fold inflated projective plane with deficiency 0 has also been called an inflated affine plane.

We want to prove the following characterization of s-fold inflated projective planes.

<u>16.1 Theorem</u> (Metsch, 1991c). Let L be a linear space, q_0 one of its points, and denote the degree of q_0 by r_0. Suppose that there exists integers $n \geq 2$ and $c \geq 3$ and a line L_0 with the following properties.

1) L_0 passes through q_0.

2) L_0 and each point of $L_0-\{q_0\}$ has degree $n+1$.

3) $v = n^2+c$ and $b \leq n^2+c+r_0-2$.

4) L_0 is parallel to at most $c-2$ lines.

5) There exists a line which is parallel to L_0.

6) Every point has degree at least $n+1$ and every line has degree at most $n+1$.

7) Every point p of $L_0-\{q_0\}$ lies on at least c lines of degree $n+1$ with equality if and only if every line through p has degree n or $n+1$.

Then L is an s-fold inflated projective plane of order n for some integer s with $1 \leq s < n$. If $D_1,\ldots,D_s$ are its main subspaces then q_0 is a point of every subspace D_j. Furthermore, if b_j is the number of lines of D_j, v_j its number of points, and r_j the degree of q in D_j, and if d is the deficiency of L, then $b_j \leq v_j+r_j-2$ for all j, and

$$2+d+(s-1)(n+1)+\sum_{j=1}^{s}(b_j-v_j-r_j) \leq 0.$$

The first step of the proof of this theorem is a reduction.

<u>16.2 Lemma</u>. In order to prove Theorem 16.1, it suffices to show that every linear space L which satisfies the hypotheses of Theorem 16.1 has a subspace $D = (A,\mathcal{L}')$ with $A\cap L_0 = \{q_0\}$ and such that some point $q \neq q_0$ has degree r_q-n in D (r_q is the degree of q in L).

<u>Proof</u>. Suppose that we have already shown that every linear space which satisfies the hypotheses of 16.1 has a subspace with the described properties.

Suppose that L satisfies the hypotheses of 16.1. Then L has a subspace $D_1 = (A_1,\mathcal{L}_1)$ with $A_1\cap L_0 = \{q_0\}$ and there is a point $q \neq q_0$ in A_1 such that q has degree r_q-n in D. We denote by b_1 the number of lines of D_1, by v_1 its number of points, and by r_1 the degree of q_0 in D_1. We smooth L by replacing D_1 by one line $H_1:=A_1$ and obtain a new linear space L' with $v' := v$ points and $b' := b+1-b_1$ lines. The point q_0 has degree $r' := r+1-r_1$ in L', and q has degree $n+1$ in L'. We have

$$b' = b+1-b_1 \leq v+r-1-b_1 < v+r-1-r_1 = v'+r'-2.$$

If L_0 is in L' parallel to some line, then also L' satisfies the hypotheses of Theorem 16.1, and it follows in the same way that L' has a subspace $D_2 = (A_2,\mathcal{L}_2)$ with $A_2 \cap L_0 = q_0$. If p is any point $\neq q_0$ of D_2 then p has degree at least $n+2$ in

L', since it has degree at least two in D_2 and because the lines px, $x \in L_0-\{q_0\}$ are not in D_2. Since q_1 has degree n+1 in L' it follows that q_1 is not in A_2. Since A_1 is a line of L', this implies that $A_1 \cap A_2 = \{q_0\}$.

Repeating this process we see that there exists an integer $s \geq 1$ and subspaces $D_1,\ldots,D_s$ of L any two of which have only the point q_0 in common and such that every line which is parallel to L is a line of one of these subspaces. We smooth L by replacing each subspace D_j by a line H_j and obtain a linear space E in which L_0 meets every line.

Let $L \neq L_0$ be a line of degree n+1 of E which meets L_0 in a point other than q_0. Since L_0 meets every other line, the Transfer Lemma implies that also L meets every other line in E. It follows that q_0 has degree n+1 in E. Consequently E has n^2+n+1 lines and can be extended to a projective plane of order n (Corollary 3.5). By definition, L is therefore an s-fold inflated projective plane of order n. Let d be the deficiency of L, let b_j be the number of lines of D_j, let v_j be its number of points, and let r_j be the degree of q_0 in D_j. Then L has $n^2+n+1-s+\sum b_j$ lines and $1+(n+1-s)n-d+\sum v_j$ points. Furthermore q_0 has degree $r = n+1-s+\sum r_j$ in L. In view of $b \leq v+r-2$, it follows that

$$(+) \qquad \sum b_j \leq \sum (v_j+r_j) - (s-1)(n+1)-d-2.$$

If we fix an index k and use $v_j \leq n+1$ and $r_j \leq b_j$ for all indices $j \neq k$ then (+) implies that $b_k \leq v_k+r_k-2$. In view of $b_j \geq r_j+1$ and $v_j \leq n+1$ for all j, equation (+) implies furthermore that $s \leq n-1$. This completes the proof of the lemma.■

In order to prove Theorem 16.1, it suffices therefore to show that every linear space satisfying its hypotheses has a subspace with the required properties. This will be shown in the rest of the section. We use the following notation. Throughout, L denotes a linear space satisfying the hypotheses of Theorem 16.1 We call a line

- *real* — if it contains a point $\neq q_0$ of degree n+1,
- *long* — if it has degree n+1, and
- *hyperideal* — if it is parallel to L_0 and if it meets every long line not passing through q_0,

A point $p \neq q_0$ will be called

real	if it has degree $n+1$,
ideal	if it has degree $> n+1$ and if the line pq_0 is not real, and
imaginary	if it has degree $> n+1$ and if the line pq_0 is real.

16.3 Lemma. Suppose that N is an n-line which meets L_0 in a real point and that N is parallel to two intersecting lines L_1 and L_2. Denote the number of lines which meet N and miss L_0 by s, and denote the degree of L_j by $n+1-d_j$. Then $d_1d_2 \geq n-s$. Furthermore one of the lines L_1 and L_2 has degree at most $n-1$.

Proof. The Parallel-Lemma in Section 1 implies that $d_1d_2 \geq n-s$. Since L_0 misses at most $c-2$ lines, we have $s \leq c-2$ so that $d_1d_2 \geq n+2-c$. Since every real point of L_0 lies on at least c long lines and because N meets L_0 in a real point, we have $c \leq n$. Hence $d_1d_2 \geq 2$ i.e. one of the lines L_1 and L_2 has degree at most $n-1$.■

16.4 Lemma. Suppose that every real point of L_0 lies on exactly c long lines. Then L has a subspace $D = (A, \mathcal{L}')$ with $A \cap L_0 = \{q_0\}$ and such that some point $q \neq q_0$ has degree $r_q - n$ in D.

Proof. By hypothesis 7), every real point of L_0 lies on c long lines and on $t := n+1-c$ lines of degree n. Fix a real point p of L_0, denote by $N_1,\dots,N_t$ the n-lines through p, and let Π_j be the set consisting of N_j and the $n-1$ lines which miss N_j and meet L_0 in a real point. Since N_j meets every long line (since $p \in N_j$), the set Π_j consists of n lines of degree n. By Lemma 16.3, the lines of Π_j are mutually parallel. If N is any n-line which meets L_0 is a real point $\neq p$ then N is parallel to a unique line N_j so N is contained in exactly one of the sets Π_j.

Let T be the set consisting of L_0 and the n^2 lines which meet L_0 in a real point, and denote by E' the incidence structure which is induced by the lines of T on the $v = n^2+c$ points of L. In E' we let the lines of Π_j meet in a new (infinite) point ∞_j, $j = 1,\dots,t$, and obtain a new structure E with n^2+n+1 points and n^2+1 lines. Every line of E has degree $n+1$ and any two different lines of E meet in a unique point. Hence E can be extended to a projective plane P of order n by adjoining n new lines $L_1,\dots,L_n$ which pass through q_0 (apply Corollary 3.5 to the structure dual to E). Because every line of T has degree at least n, the lines of $T-\{L_0\}$ cover exactly those pairs (x,y) of points of L for which x and y are not contained in the same line L_j. Hence every line of L which has two points in common with some line L_j is already contained in L_j. If the line L_j contains at least two lines of L then the lines of L which are contained in L_j induce therefore a subspace of L. Consequently L is an s-fold inflated projective plane of order n for some integer s. We have $s \geq 1$ since there is a line which is parallel to L_0. Every main subspace of L satisfies the conclusion of this lemma.■

<u>16.5 Lemma</u>. Suppose that L is a long line and that p is a point outside of L_0 and L. Let $H_1,\ldots,H_s$ be the lines through p which are parallel to L_0 ($s = 0$ is allowed). Then L meets exactly one of the lines $pq_0, H_1,\ldots,H_s$.

<u>Proof</u>. Since L is long it meets every line px where x is a real point of L_0. Since these are n lines, the assertion follows.■

The rest of the proof is divided into two parts.

A. <u>Every ideal point lies on a hyperideal line.</u>

Throughout in this subsection, we assume that every ideal point lies on a hyperideal line. Notice that this corresponds to the case in Theorem 16.1 in which the main subspaces D_j are near-pencils and such that the "long" line of D_j is disjoint to L_0.

Every line which is parallel to L_0 and which is not hyperideal will be called *imaginary*. By Q_0 we denote the set consisting of the real lines passing through q_0 and of the hyperideal lines. Furthermore, for every n-line N with $q_0 \notin N$ and $N \cap L_0 \neq \emptyset$ we denote by π_N the set consisting of N and all real and hyperideal lines which miss N. Notice that a line X which is parallel to N is in π_N if and only if it meets L_0 in a real point or if it is in Q_0.

<u>16.6 Lemma</u>. a) $|Q_0| = n+1$, and every point $\neq q_0$ lies on a unique line of Q_0.
b) A hyperideal line contains only ideal points.

<u>Proof</u>. a) Let p be any point $\neq q_0$. If p is a point of L_0, then L_0 is the only line of Q_0 through p. Assume now that $p_0 \notin L_0$. If p is real or imaginary then the line xq_0 is real and in Q_0. If p is ideal then p lies on a hyperideal line, which is in Q_0. Therefore p lies on at least one line of Q_0. We have two show that this

line is unique. Since every real point of L_0 lies on $c \geq 3$ long lines, there is a long line L with $p, q_0 \notin L$. Let $G_1,\dots,G_s$ be the lines through p which miss L_0. Then Lemma 16.5 shows that L meets exactly one of the lines $pq_0, G_1,\dots,G_s$. Since L meets every real line (since real lines have a real point) and every hyperideal line (by definition of *hyperideal*), this shows that at most one of the lines $pq_0, G_1,\dots,G_s$ is in Q_0.

b) Let p be any point of a hyperideal line H. Then H is the unique line of Q_0 through p. Therefore the line pq_0 is not in Q_0, which means that pq_0 is not real. By definition, p is an ideal point.■

16.7 Lemma. Let N be a real n-line which meets L_0 in a real point (so $q_0 \notin N$).

a) $|\Pi_N| = n+1$ and $|\Pi \cap Q_0| = 1$.

b) Let p be any point with $p \neq q_0$ and $p \notin N$. Then p lies on at least one line of Π_N. Furthermore, p lies on (at least) two lines of Π_N, if and only if

(1) the line pq_0 is not real and meets N, or

(2) p lies on an imaginary line which meets N.

Proof. a) Since $|N| = n$, 16.6 shows that Π_N and Q_0 have a unique line L in common. By definition of Π_N and Q_0, every line of $\Pi_N - \{L\}$ meets L_0 in a real point. Since every real point of L_0 lies on a unique line of Π_N, this shows that $|\Pi_N| = n+1$.

b) If $p \in L_0 - \{q_0\}$, then p lies on a unique line of Π_N and (1) and (2) are not satisfied. Assume now that $p \in L_0$ and let s be the number of lines through p which meet L_0 and miss N. Then $s-1$ is the number of lines through p which meet N and miss L_0 (Transfer Lemma). This implies that $s \geq 1$. We consider three cases.

Case 1. pq_0 *is a real line.* Then the s lines through p which miss N and meet L_0 are in Π_N. Since the line pq_0 is in Q_0, Lemma 16.6 shows that p does not lie on a hyperideal line. Hence, the $s-1$ lines through p which meet N and miss L_0 are imaginary. Thus, (2) is fulfilled if and only if p lies on at least two lines of Π_N.

Case 2. pq_0 *is not real and the hyperideal line through* p *meets the line* N. Since pq_0 is not real, Lemma 16.6 a) shows that p lies on a unique hyperideal line X. We have assumed that X meets N. Since $s-1$ is the number of lines which meet N and miss L_0, we have $s \geq 2$ and p lies on $s-2$ imaginary lines which meet N.

If pq_0 misses N, then p lies on s−1 lines of Π_N, since pq_0 is not in Π_N. Since s−2 is the number of imaginary lines through p which meet N, it follows that p lies on at least one line of Π_N with equality if and only if (2) is not fulfilled.

If pq_0 meets N, then (1) is fulfilled and the $s \geq 2$ lines through p which meet L_0 and miss N are all in Π_N.

Case 3. pq_0 *is not real and the hyperideal line through* p *misses* N. If X is the hyperideal line through p, then X is in Π_N. Hence, p lies on at least one line of Π_N. Now the s−1 lines through p which meet N and miss L_0 are imaginary. Furthermore, if pq_0 meets N, then p lies on $s+|\{X\}|\geq 2$ lines of Π_N, and if pq_0 misses N, then p lies on $s-1+|\{X\}|$ lines of Π_N. This proves part c) in this case.■

<u>16.8 Lemma</u>. Suppose that N is an n-line which meets L_0 in a real point. Then every n-line of Π_N misses every other line of Π_N.

<u>Proof</u>. Assume by way of contradiction that Π_N contains an n-line N' which meets a second line H of Π_N and denote the degree of H by n+1−d.

By Lemma 16.7, every point other than q_0 lies on at least one line of Π_N. Furthermore the point of intersection of N' and H lies on two lines of Π_N. If we count incident point-line pairs (p,L) with $L \in \Pi_N$ then we obtain

$$n^2+c = (v-1)+1 \leq \sum_{X\in\Pi_N} k_X.$$

Since every line of Π_N has degree at most n (this follows from the fact that N contains a real point of L_0), we conclude that

$$n^2+c \leq |\Pi_N-\{H\}|n+k_H = n^2+k_H = n^2+n+1-d$$

so that $d \leq n+1-c$. On the other hand, if s is the number of lines which meet N and miss L_0 then $d \geq n-s$ (Lemma 16.3). Hence $s \geq c-1$. But L_0 is parallel to at most c−2 lines (hypothesis 4) in Theorem 16.1), a contradiction.■

<u>16.9 Lemma</u>. Suppose that $N_1,\dots,N_u$ are mutually parallel n-lines which meet L_0 in a real point and let p be a point outside of L_0. Suppose that p lies on two lines H_1 and H_2 which are real or hyperideal and which are parallel to every line N_j. Then $u \leq c-1$.

Proof. Let Π_j be the set consisting of N_j and the real and hyperideal lines which miss N_j. Since p lies on two lines of Π_j, Lemma 16.7 shows that p lies on a line K_j, $j = 1,\dots,u$, which meets N_j such that

1) $q_0 \in K_j$ and K_j is not real, or

2) K_j is imaginary.

The same lemma shows furthermore that each point q of K_j with $q \neq q_0$ and $q \notin N_j$ lies on at least two lines of Π_j. It follows that the lines K_j are distinct (if for example K_1 and K_2 were equal, then the point $K_1 \cap N_2$ would lie on a line $X \neq N_2$ of Π_1, which contradicts Lemma 16.8). Hence, $K_j = pq_0$ for at most one value for j so that there are at least u−1 lines K_j satisfying 2). Since L_0 misses every imaginary line and because L_0 misses at most c−2 lines, it follows that $u-1 \le c-2$.■

16.10 Lemma. Suppose that N is an n-line which meets L_0 in a real point. Then the lines of Π_N are mutually parallel.

Proof. Assume to the contrary that two lines L_1 and L_2 of Π_N meet, and denote by u+1 the number of n-lines in Π_N. By Lemma 16.8, L_1 and L_2 have degree at most n−1. Using the technique of the proof of Lemma 16.8, we see that

$$n^2+c \le \sum_{X\in\Pi_N} k_X \le (|\Pi_N|-u-3)(n-1)+(u+1)n+|L_1|+|L_2|$$

$$= n^2-2n+u+2+|L_1|+|L_2| \le n^2+u.$$

Hence $u \ge c$.

Since n of the n+1 lines of Π_N meet L_0 in a real point, at least u of the n-lines of Π_N meet L_0 in a real point. By Lemma 16.8, these n-lines are mutually parallel and miss L_1 and L_2. Since $u \ge c$, Lemma 16.9 shows that this is not possible.■

16.11 Lemma. a) $\Pi_N = \Pi_{N'}$ or $|\Pi_N \cap \Pi_{N'}| = 1$ for any n-lines N and N' which meet L_0 in a real point.

Proof. If N and N' are parallel then Lemma 16.10 shows that $\Pi_N = \Pi_{N'}$. If N' meets N, then N' meets n of the lines of Π_N (Lemma 16.10 and 16.7) so that $|\Pi_N \cap \Pi_{N'}| = 1$.■

16.12 Lemma. a) Suppose that H is a hyperideal line. Then every line xq_0 with $x \in H$ has degree 2, and H meets no imaginary line.

b) If R is a real line then R meets no imaginary line.

c) Every line parallel to L_0 is hyperideal.

Proof. a) Denote the degree of H by n−d and let T be the set consisting of the nd lines $\neq L_0$ which miss H and meet L_0 in a real point. Furthermore let $\pi_1,\dots,\pi_e$ be the sets π_N where N is an n-line of T. Then $\pi_j \subseteq T\cup\{H\}$ for all j and $\pi_j\cap\pi_k=\{H\}$ for all distinct j and k (Lemma 16.10, 16.7 a), and 16.11). Set $T' := T-(\pi_1\cup\dots\cup\pi_e)$. Then $|T'|=(d-e)n$, which implies that $e \leq d$. Let F be the set of points p with $p \neq q_0$ and $p \notin H$, and let t_p be the number of lines of T' through p for every point $p \in F$. Since every line of T' is contained in F, we have

$$\sum_{p\in F} t_p = \sum_{X\in T'} k_X$$

Every point p of F lies on at least d+1 lines which meet L_0 and miss H. One of these lines may be the line pq_0 and e of these lines are in $\pi_1\cup\dots\cup\pi_e$ so at least d−e of them are in T'. Consequently $t_p \geq d-e$ with equality if and only if the line pq_0 is parallel to H and if d+1 is the number of lines through p which miss H and meet L_0.

On the other hand, every line of T has degree at most n (since H is hyperideal) and every line of T' has degree at most n−1 (since the n-lines of T induce the sets π_j). It follows that

$$(v-1-k_H)(d-e) \leq \sum_{p\in F} t_p \leq \sum_{X\in T'} k_X \leq |T'|(n-1) = (d-e)n(n-1).$$

In view of $v-1-k_H > n^2-k_H \geq n^2-n$, we obtain $d = e$ and $t_p = d-e = 0$ for every point p of F.

It follows that every line pq_0 with $p \in F$ is parallel to H, which implies that the lines hq_0, $h \in H$, have degree 2. For a point $p \in F$ with $p \notin L_0$, the equation $t_p = d-e = 0$ implies furthermore that p lies on exactly d+1 lines which meet L_0 and miss H. The Transfer Lemma shows therefore that p lies on no line which meets H and misses L_0. Consequently H meets no (imaginary) line parallel to L_0.

b) Put $|R| = n+1-d$. As in part a), let T be the set consisting of the nd lines which meet L_0 and miss R, let $\pi_1,\dots,\pi_e$ be the sets π_N induced by the n-lines of T, and let T' be the set consisting of the (d−e)n lines of T which are not in one of the sets π_j, $j = 1,\dots,e$. Then every point p outside of R lies on at least d−e lines of T' with equality if and only if p lies on no line which meets R and misses L_0, and every line of T' has degree at most n−1. As in part a), this implies that d = e and shows that every point outside of R lies on exactly d lines of T, and that R meets no line which is parallel to L_0.

c) Every point outside of L_0 lies on a real or hyperideal line of Q_0 (Lemma 16.6). Part c) is therefore an immediate consequence of part a) and b).■

<u>16.13 Lemma</u>. L has a subspace $D = (A,\mathcal{L}')$ with $A \cap L_0 = \{q_0\}$ and such that some point $q \neq q_0$ has degree $r_q - n$ in D.

<u>Proof</u>. Since L satisfies the hypotheses of Theorem 16.1, there exists a line H which is parallel to L_0. In view of Lemma 16.12 c), H is hyperideal. By Lemma 16.12 a), the set $A := H \cup \{q_0\}$ together with the line H and the 2-lines q_0h, $h \in H$, is a subspace of L. Let q be any point of H. Since H meets no other line which is parallel to L_0 (Lemma 16.12 a) and Lemma 16.6), the point q has degree n+2 in L. In the subspace, which is a near-pencil, q has degree $2 = r_q - n$.■

<u>Remark</u>. It is easy to see that L not only has a subspace but that L is actually an s-fold inflated projective plane. However, in order to prove Theorem 16.1, we only need to know that there is one subspace (see Lemma 16.2).

B. <u>Some ideal point does not lie on any hyperideal line.</u>

From now on, we assume that there exists an ideal point which does not lie on any hyperideal line. Recall that we have to show that L has a subspace $D = (A,\mathcal{L}')$ with $A \cap L_0 = \{q_0\}$ and such that some point $q \neq q_0$ has degree $r_q - n$ in D. A line X which misses L_0 will be called *ideal* if it meets some but not all of the long lines L with $q_0 \notin L$. Recall that a hyperideal line is a line which misses L_0 and which meets every long line L with $q_0 \notin L$.

We fix an ideal point q which lies not on any hyperideal line and denote its degree by n+2+a. We set $Z := qq_0$ and denote the a+1 lines through q which are parallel to L_0 by $Z_0,\dots,Z_a$. We denote the degree of Z by z+1 and the degree of Z_j

by z_j+1. Furthermore we define A to be the set of points which lie on Z or one of the lines Z_j. We may assume that $z_j \geq z_k$ for $k \geq j$. We may also assume that we have chosen q in such a way that the following property holds.

16.14 Property. Suppose that q' is an ideal point which does not lie on any hyperideal line. Then

a) q' has degree at least n+2+a.

b) If q' has degree n+2+a and if X is a longest line through q' which misses L_0, then $|X| \leq |Z_0|$.

We shall show that the lines contained in A induce a linear space D on A. In view of Lemma 16.4, we may assume that L_0 has a real point which lies on c+1 long lines.

16.15 Lemma. a) Every long line contains a unique point of A. In particular every real point of L_0 lies on at most $|A|$ long lines. At least c+1 points of A lie on a long line.

b) $c+1 \leq |A| = 1+z+\sum_{j=0}^{a} z_j$.

c) $z \leq c-2$.

Proof. a) follows from Lemma 16.5, the definition of A, and the assumption that some real point of L_0 lies on at least c+1 long lines (see the paragraph preceding this lemma).

b) By definition we have $|A| = 1+z+\sum z_j$. Part a) implies that $|A| \geq c+1$.

c) The point q is ideal; by definition this means that every point of $Z-\{q_0\}$ has degree at least n+2. Therefore Z meets at least z lines missing L_0. Since L_0 is parallel to at most c−2 lines, it follows that $z \leq c-2$.■

16.16 Lemma. Suppose that L is a long line and that I is an ideal line which meets L in a point p. Let x be any point of $I-\{p\}$ and denote the line through x and q_0 by H.

a) L meets L_0 in a real point.

b) The point x is ideal and has degree at least n+2+a.

c) The point x lies not on any hyperideal line.

d) The line H is parallel to L and therefore not real.

e) If m is the number of lines which miss the lines L_0, I, and H then $(|I|-1)a+(|H|-1) \leq c-2-m$.

f) If there exists a long line L' which meets H in a point $\neq q_0, x$ then $(|H|-2)(a+1)+(|I|-1) \leq c-2$.

Proof. By the Transfer Lemma, x lies on a line X which meets L_0 and misses L. Since q_0 is the only non-real point of L_0, X passes through q_0. Consequently $X = H$. It follows that $q_0 \notin L$ and that H has no real point. This proves a) and d) and shows that x is an ideal point.

By definition, every hyperideal line meets L. Since L meets I, Lemma 16.5 shows therefore that x lies not on any hyperideal line. It follows that x has degree at least $n+2+a$ (Property 16.14).

The same argument shows that every point of $I-\{p\}$ has degree at least $n+2+a$. Consequently, there are at least $(|I|-1)a$ lines $\neq I$ which are parallel to L_0 and meet I in a point $\neq p$. By Lemma 16.5, all these lines are also parallel to L. By the Transfer Lemma, every point of $H-\{q_0\}$ lies on a line which meets L and misses L_0. Hence there exists at least $(|I|-1)a+(|H|-1)$ lines which miss L_0 and meet H or I. Since L_0 is parallel to at most $c-2$ lines, this proves e).

In order to prove part f), suppose that there is a long line L' which meets H in a point $s \neq q_0, x$. Let y be any point of $H-\{q_0, s\}$. Then the $n+1$ lines through y which meet L_0 meet also L', since each of these $n+1$ lines is either H or meets L_0 in a real point. Therefore every line through y which misses L_0 misses also L'. Since L' meets every hyperideal line (by definition of *hyperideal*), it follows that y does not lie on a hyperideal line. Hence y has degree at least $n+2+a$ (Property 16.14) and lies on $a+1$ lines which miss L_0 and L'. Consequently, there are $(|H|-2)(a+1)$ lines which meet H and miss L_0 and L'. One of these lines is I. The Transfer Lemma shows that there exists at least $|I|-1$ lines which meet L' and miss L. Hence there are also at least $|I|-1$ lines which meet L' and miss L_0 (a point of L' which lies on s lines missing L lies also on s lines missing L_0). Since L_0 is parallel to at most $c-2$ lines, this proves f).■

16.17 Lemma. Suppose that L_1 and L_2 are long lines and that I_1 and I_2 are two lines missing L_0. Suppose furthermore that I_1 and I_2 meet in a point p and that L_j and I_j meet in a point $p_j \neq p$, $j = 1,2$. Then $(|I_1|-1)a+(|I_2|-1) \leq c-2$.

Proof. By Lemma 16.16 b), every point of $I_1-\{p_1\}$ has degree at least n+2+a. Therefore there are at least $(|I_1|-1)a$ lines which miss L_0 and meet I_1 in a point $\neq p_1$. By Lemma 16.5 these lines are also parallel to L_1.

The Transfer Lemma shows that there are $|I_2|-1$ lines which meet L_1 and miss L_2. Since every point $\neq L_1 \cap L_0$, $L_1 \cap L_2$ of L_1 lies on the same number of lines which miss L_0 or L_2 and because L_1 and L_0 meet in a point of degree n+1 (Lemma 16.16 a)), it follows that there are also $|I_2|-1$ lines which meet L_1 and miss L_0. Since L_0 is parallel to at most c−2 lines, the assertion follows.■

16.18 Lemma. a) $a \geq 1$.

b) $z_j a+z \leq c-2$ for all $j \in \{0,\ldots,a\}$.

c) Every line Z_j meets a long line in a point other than q.

d) $z_0 a+z_1 \leq c-2$.

e) $\sum_{j=0}^{a} z_j \leq c-2$ and $\sum_{j=0}^{a} z_j \leq c-2-z+z_1$.

f) $c+1 \leq |A| \leq c-1+z$ and $c+1 \leq |A| \leq c-1+z_1$.

g) The point q lies on at least $n+1-z_1 > 0$ long lines.

Proof. Since the ideal point q has degree n+2+a, we have $a \geq 0$. Assume by way of contradiction that $a = 0$. Then Z and Z_0 are the only non-real lines through q_0. Lemma 16.15 shows that $c \leq z+z_0$ and $z \leq c-2$. Fix a real point p of L_0. Since p lies on at least c long lines, p lies on a long line L missing Z. By Lemma 16.15 a), L meets Z_0. Since Z_0 is an ideal line, there is also a long L' not passing through q_0 which misses Z_0. By Lemma 16.15 a), the line L' meets Z. Lemma 16.16 f) implies that $(z-1)(a+1)+z_0 \leq c-2$, i.e. $z+z_0 \leq c-1$, a contradiction. Thus $a \geq 1$.

Let M be the set of indices j such that the line Z_j meets a long line in a point other than q. Lemma 16.15 a) shows that

$$c+1 \leq 1+z+\sum_{j \in M} z_j$$

By Lemma 16.16 e) we have $z_j a+z \leq c-2$ for all $j \in M$. It follows that

$$c+1 \le |A'| \le 1+z+|M|\frac{c-2-z}{a} \quad ,$$

which implies that $|M| > a$, i.e. $M = \{0,\dots,a\}$. This proves b) and c).

Since there are long lines G_j, $j = 0,1$, which meet Z_j in a point other than q, Part d) follows from Lemma 16.17. In view of $z_0 \ge z_j$ for all j, we obtain $\sum z_j \le z_0 a+z_1 \le c-2$ from d) and $\sum z_j \le z_0 a+z_1 \le c-2-z+z_1$ from b). Part f) is an immediate consequence of e) and Lemma 16.15 b).

It remains to prove part g). Let α be the number of long lines through q. Counting the number of points from q, we find

$$v \le |A|+(n-\alpha)(n-1)+\alpha n$$

and hence $\alpha \ge v+n-n^2-|A| = n+c-|A| \ge n+1-z_1$. In view of $z_1 \le c-2$ (part b)) and $c \le n+1$ (since a real point of L_0 lies on at least c long lines) we have $n+1-z_1 > 0$.■

<u>16.19 Lemma</u>. a) No hyperideal line meets A.

b) Every point of $A-\{q_0\}$ is ideal and has degree at least $n+2+a$.

c) $z(a+1) \le c-2$.

<u>Proof</u>. a) By Lemma 16.18 g), there exists a long line L passing through q. Let U be any hyperideal line. Because q lies not on any hyperideal line (q was chosen in this way), we have $q \notin U$. Since U meets L (by definition of *hyperideal*), Lemma 16.5 shows that U does meet any of the lines $Z, Z_0, \dots, Z_a$. Hence $A \cap U = \emptyset$.

b) Every point of $Z-\{q_0\}$ is ideal, since q is ideal. Since q lies on a long line, Lemma 16.16 b) shows that also every point of $A-Z$ is ideal. From part a) and Property 16.14 it follows that every point of A has degree at least $n+2+a$.

c) Since every point of $Z-\{q_0\}$ lies on at least $a+1$ lines parallel to L_0 and because L_0 is parallel to at most $c-2$ lines, we have $z(a+1) \le c-2$.■

<u>16.20 Lemma</u>. There exists a point $\neq q$ of degree $n+2+a$ on Z_0.

<u>Proof</u>. By Lemma 16.18 c), there exists a long line L meeting Z_0 in a point $p \neq q$. Assume by way of contradiction that every point $\neq q$ of Z_0 has degree at least $n+3+a$, and let α be the number of lines which are parallel to L_0 and which meet Z_0 in a point $\neq p$. Then

$$\alpha = \sum_{\substack{x\in Z_0\\ x\neq p}} (r_x-n-2) \geq z_0(a+1)-1.$$

These α lines are also parallel to L (Lemma 16.5). Since there are z lines which meet Z and L and miss L_0 (this follows from the Transfer Lemma, since Z meets L_0 and misses L) and since L_0 is parallel to at most c−2 lines, it follows that $z_0(a+1)-1+z \leq c-2$. In view of $z_0 \geq z_j$ for all j, it follows that $1+z+\sum z_j \leq c$, contradicting Lemma 16.15.■

From now on q' denotes a point $\neq q$ of degree n+2+a of Z_0, and $Y_1,\ldots,Y_a$ denote the lines $\neq Z_0$ through q' which miss L_0 such that $|Y_j| \geq |Y_k|$ for $j \leq k$. We set $Y_0 := Z_0$, $Y := q'q_0$, we denote the degree of Y by y+1 and the degree of Y_j by y_j+1. By Property 16.14, we have $z_0 = y_0 \geq y_1 \geq \ldots \geq y_a$.

Since q' has the same properties as q, everything what we have proved for the lines Z, Z_j and the values z, z_j holds also for the lines Y, Y_j and the values y, y_j. It is no loss of generality to assume that $y_1 \geq z_1$ and we shall do so.

<u>16.21 Lemma</u>. Let G be a long line which meets A in a point $p' \neq q_0$, and suppose that p' lies on a non-real line H which has a point p outside of A. Then there are at least 2a+1 lines which are parallel to L_0 and which have no point in common with A.

<u>Proof</u>. Since A consists of the points of the non-real lines through q, we have $p' \neq q$. By Lemma 16.18 g), the point q lies on $n+1-z_1 > 0$ long lines. Let L be any long line through q. Then $L \cap H = \emptyset$ by Lemma 16.5. By the Transfer Lemma, p lies on a line X_L which misses G and meets L. The line X_L is not real, since it is parallel to G. Furthermore X_L has no point in common with $A-\{q_0\}$ (Lemma 16.5).

Assume by way of contradiction that all the lines X_L, L a long line through q, are distinct. Then p lies on $n+1-z_1$ non-real lines. One of these lines may be pq_0 but the others are parallel to L_0. Hence p lies on $n-z_1$ lines which miss L_0 and have no point in common with A. But consider the line Z_1 of degree z_1+1, which is parallel to L_0. Each of its points lies on another line which is parallel to L_0 and this gives z_1+2 lines which are parallel to L_0 and have a point in A. Since L_0 misses at most $c-2 < n$ lines, this is a contradiction.

Consequently q lies on two long lines L and L' with $X_L = X_{L'}$. Let x be the point of intersection of L and X_L.

Then x is an ideal point. (For: If x were not an ideal point then the line $R := xq_0$ would be real. Hence $R \neq X_L$ and R meets L'. But now R and X_L pass through x and meet L'. Lemma 16.5 shows that this is not possible.)

The point x lies not on any hyperideal line U. (For: U would meet L' and G so that $U \neq X_L$. This time U and X_L pass through x and meet L'. In view of Lemma 16.5, this is not possible.)

In view of Property 16.14, x has therefore degree at least n+2+a. Hence x lies on a+1 lines which miss L_0. Because the long line L passes through q and x, these a+1 lines have no point in A (Lemma 16.5). Similarly, the point x' in which L' and $X_L = X_{L'}$ meet lies on a+1 lines which miss L_0 and have no point in common with A. Hence there are at least 2(a+1)−1 lines which miss L_0 and A.■

<u>16.22 Lemma</u>. $z_1 \leq a+1$.

<u>Proof</u>. Assume to the contrary that $z_1 \geq a+2$. Then also $y_1 \geq a+2$ (see the paragraph preceding Lemma 16.21) so that there is a point on Y_1 which lies not on one of the lines $Z, Z_0, Z_1, \ldots, Z_a$. Lemma 16.21 shows therefore that there are 2a+1 lines which miss L_0 and A. Since Z_0 meets a long line L in a point other than q (Lemma 16.18 c)), it follows from Lemma 16.16 e) (applied with $p = q$, $I = Z_0$, and $H = Z$) that $z_0 a + z \leq c-2-(2a+1)$. It follows that

$$|A| \leq 1+z+\sum z_j \leq 1+z+az_0+z_1 \leq c-2-2a+z_1.$$

Let α be the number of long lines through q and count the number of points from q to obtain

$$n^2+c = v \leq |A|+(n-\alpha)(n-1)+\alpha n = n^2-n+|A|+\alpha.$$

We obtain $\alpha \geq n+c-|A| \geq n+2a+2-z_1$. However if L is any long line through q then L misses Y_1 (since Z_0 is the non-real line through q' which meets L, see Lemma 16.5) and q lies therefore on at most $r_q - y_1 = n+2+a-y_1$ long lines. We obtain $y_1 \leq z_1 - a$. This is a contradiction, since $y_1 \geq z_1$ (see the paragraph preceding Lemma 16.21) and $a \geq 1$ (Lemma 16.18).■

<u>16.23 Lemma</u>. $z_0 = z_1$.

<u>Proof</u>. Set $\sigma = z_0 - z_1$. Then $z_j \le z_0 - \sigma$ for all $j \ge 1$. Hence, $\sum z_j \le z_0 a + z_1 - (a-1)\sigma$. In view of $z_0 a + z \le c-2$ (Lemma 16.18 b)), we conclude that

$$c+1 \le |A| = 1+z+\sum z_j \le c-1+z_1-(a-1)\sigma \le c+a-(a-1)\sigma.$$

Hence $(a-1)\sigma \le a-1$.

If $(a-1)\sigma < a-1$ then $\sigma = 0$ and we are through. Let us therefore suppose that $(a-1)\sigma = a-1$. Then we obtain equality in all the above inequalities. Consequently $|A| = c+1$, $z_0 a + z = c-2$, $z_1 = a+1$, and $z_0 = a+1+\sigma$.

Consider a real point x of L_0. By Lemma 16.15, each long line through x meets A in a unique point. Since $|A| = c+1$ and because x lies on at least c long lines, it follows that every line through x meets A in at most one point. Hence, every line which meets A in two points is either parallel to L_0 or passes through q_0. Hence, if p is a point of A with $p \neq q_0$ and $p \notin Z_0$, then p lies on n lines which meet L_0 in a real point and on $|Z_0| = z_0+1 = a+2+\sigma$ lines which meet A in a second point (which is on Z_0) so that p has degree at least $n+1+z_0 = n+2+a+\sigma$.

It follows that each point of $Z_1 - \{q\}$ lies on at least $a+\sigma$ lines $\neq Z_1$ which miss L_0. Since q lies on $a+1$ lines parallel to L_0 and because L_0 misses at most $c-2$ lines, we obtain $z_1(a+\sigma)+a+1 \le c-2$. In view of $c-2 = z_0 a + z$ and $z_0 = z_1 + \sigma$, it follows that $z \ge \sigma(z_1 - a)+a+1 = a+1+\sigma$.

Counting the number of lines parallel to L_0 which meet Z shows in the same way that $(z-1)(a+1+\sigma)+a+1 \le c-2 = z_0 a + z = (a+1+\sigma)a+z$, i.e $(z-a-1)(a+\sigma) \le 0$. Hence $z \le a+1$. Since we also have proved that $z \ge a+1+\sigma$, we see that $\sigma = 0$.■

<u>16.24 Lemma</u>. a) There are at most $a-1$ lines which are parallel to L_0 and Z_0 with equality only if every point of A lies on a long line. The same statement holds for the lines L_0 and Z_1.

b) We have $c \le a^2+3a+3$.

<u>Proof</u>. In view of $z_j \le z_0$, Lemma 16.15 shows that $c \le |A|-1 \le z+(a+1)z_0$. In view of $z(a+1) \le c-2$ (Lemma 16.19), we obtain

$$\frac{(c-2)a}{a+1} \le c-2-z \le (a+1)z_0 - 2.$$

Since $z_0 \le a+1$ (Lemma 16.23 and 16.22), we have $(a+1)z_0-2 < a(a+2)$. It follows that $c-2 < (a+2)(a+1)$, so that $z \le (c-2)/(a+1) < a+2$, i.e. $z \le a+1$.

The line Z_0 meets at least $a(z_0+1)$ other lines which are parallel to L_0. Hence there are at most $c-2-[a(z_0+1)+1]$ lines which miss Z_0 and L_0. Since $z_0 \le a+1$, we have $a(z_0+1)+1 \ge (a+1)z_0$. It follows that

$$c-2-[a(z_0+1)+1] \le c-2-[\frac{(c-2)a}{a+1}+2] = \frac{c-2}{a+1}-2 \le \frac{a^2+3a+1}{a+1}-2 < a.$$

Hence there are at most $a-1$ lines which miss Z_0 and L_0.

Now assume that there are $a-1$ lines missing L_0 and Z_0. Then we obtain $a(z_0+1)+1 = (a+1)z_0$, which implies that $z_0 = a+1$, and $c-2-[a(z_0+1)+1] = a-1$, which implies that $c = a(z_0+1)+a+2 = (a+1)(a+2)$. Now $c \le |A|-1 \le z+(a+1)z_0$ and $z \le a+1$ imply that $c = |A|-1$. Lemma 16.15 shows that each point of A lies on a long line.

Since Z_0 and Z_1 have the same degree (Lemma 16.23), the same argument shows that there are at most $a-1$ lines missing L_0 and Z_1 with equality only if each point of A lies on a long line.■

<u>16.25 Lemma</u>. Suppose that p is a point of $A-\{q_0\}$ which lies on a long line L. Then every non-real line through p is contained in A.

<u>Proof</u>. Since there are at most $a-1$ lines which are parallel to L_0 and Z_0, this follows from Lemma 16.21.■

<u>16.26 Lemma</u>. Every point of A lies on a long line.

<u>Proof</u>. Let A_1 be the set of points $\ne q_0$ of A which lie on a long line and put $A_2 = A-A_1$. Lemma 16.15 a) shows that $|A_1 \cup \{q_0\}| \ge c+1$, i.e. $|A_1| \ge c$. By Lemma 16.18 f) we have $|A| \le c-1+z_1 \le c+a$. Hence $|A_2| \le a$.

Assume by way of contradiction that A_2 contains a point $p \ne q_0$, and let S be the set of non-real lines passing through p (notice that the line pq_0 belongs to S since it misses every long line passing through q (Lemma 16.5)). For every line I of S we denote by α_I the number of points of I which are in $A-\{q_0,p\}$, and by β_I the number of points of I which are not in A. Since the n real lines through p have degree at most n, we have

$$n^2+c = v \le 1+n^2-n+\sum_{I\in S}(|I|-1) = 1+n^2-n+|\{q_0\}|+\sum_{I\in S}\alpha_I+\sum_{I\in S}\beta_I.$$

Let S_1 be the set of lines of S which have a point in A_1 and put $S_2 = S-S_1$. By Lemma 16.25 we have $\beta_I = 0$ for all $I \in S_1$. It follows that

$$n+c-2 \le \sum_{I\in S}\alpha_I + \sum_{I\in S}\beta_I = \sum_{I\in S_1}\alpha_I + \sum_{I\in S_2}\alpha_I + \sum_{I\in S_2}\beta_I$$

$$\le \sum_{I\in S_1}\alpha_I + |A_2-\{p\}| + \sum_{I\in S_2}\beta_I \le \sum_{I\in S_1}\alpha_I + \sum_{I\in S_2}\beta_I + a-1.$$

Consider a line I of S_1. By lemma 16.19 b), the α_I+1 points of $I\cap(A-\{q_0\})$ have degree at least $n+2+a$. Hence I meets at least $(\alpha_I+1)a$ lines which are parallel to L_0. Therefore $(\alpha_I+1)a \le c-2$.

Now consider a line I of S_2. For each of the β_I points x of I which are not in A, the line qx meets L_0 and is real. However it is not long, since it meets the lines pq and px (Lemma 16.5). Because q lies on n real lines and since $n+1-z_1$ of these are long (Lemma 16.18 g)), it follows $\beta_I \le z_1-1$. By Lemma 16.18 b) we have $z_1a \le c-2$. Hence $(\beta_I+1)a \le c-2$.

It follows that

$$n+c-1-a \le |S_1 \cup S_2|\frac{c-2-a}{a} \le |S|\frac{c-2-a}{a}.$$

Assume that $|S| \le 2a+1$. Then $(n+c-1-a)a \le (2a+1)(c-2-a)$ which implies that $na+a^2+4a+2 \le (a+1)c$. Since $c \le a^2+3a+3 \le a^2+4a+2$ (Lemma 6.24 and Lemma 16.18 a)), we obtain $an \le ac$, i.e. $n \le c$. Since we assume that some real point of L_0 lies on at least $c+1$ long lines (see the paragraph preceding Lemma 6.15), it follows that some real point lies only on long lines. This is not possible, since $p \in A_2$ does not lie on any long line.

Hence p lies on at least $2a+2$ non-real lines. The following argument, which we shall use in conjunction with Lemma 6.24, shows that we may assume that p lies not on the line Z_0. Then one of the non-real lines through p passes through q_0 and some of them meet Z_0. However at least $2a+2-(1+|Z_0|)=2a+2-(2+z_0) \ge a-1$ of the non-real lines through p are parallel to L_0 and Z_0. Now Lemma 16.24 shows that every point of A lies on a long line, a contradiction since $p \in A_2$ is a point not lying on any long line.■

16.27 Lemma. Any line which contains two points of A is contained in A.

Proof. Let R be any line which has two points q_1 and q_2 in A. We want to show that R is not a real line. Then Lemma 16.25 and 16.26 show that R is contained in A.

Assume to the contrary that R is a real line. Let L_j be a long line through q_j, $j = 1,2$, and let L be any long line through q. Since R can not pass through q, the lines qq_1 and qq_2 are distinct. Since L_2 meets A only in q_2 (Lemma 16.15), the line qq_1 meets L and misses L_2. By the Transfer Lemma, q_1 lies on a line I which meets L_2 and misses L. Since R is real, we have $I \neq R$. Consequently the point $I \cap L_2$ is not in A (q_2 is the only point of L_2 in A). Thus I is a non-real line which contains the point q_1 of A and the point $L_2 \cap I$ outside of A. Lemma 16.25 shows that $q_1 = q_0$. But we can show in the same way that $q_2 = q_0$ and this is not possible.■

Lemma 16.27 shows that the lines which are contained in A induce a linear space D of A. Since q has degree n+2+a in L and a+2, the subspaces D has the properties required in Lemma 16.2. This completes the proof of Theorem 16.1.

17. The Dowling Wilson Conjecture

The main result in this section is the classification of all linear spaces which satisfy $b \le v+r_q-2$ for some point q. This result will be used to prove the Dowling-Wilson Conjecture (Corollary 17.3) and to classify all Dowling-Wilson spaces (see Definition 17.4). Before we state the result, we need some terminology.

Suppose that L is a linear space. Then we can obtain a new linear space from L in the following way. We adjoin a new point and connect it to every old point by a line of degree 2. We call the obtained linear space a *simple extension of type 1* of L. We may also adjoin a new point to some line L of L and connect it to every point outside of L by a line of degree 2. In this case we call the obtained space a *simple extension of type 2* of L. In both cases the new point is called the *special point* of the extension.

17.1 Theorem (Metsch, 1991c). Suppose that L is a linear space satisfying $b \le v+r_q-2$ for some point q. Then one of the following cases occurs.

1) L can be obtained from a projective plane of order $n := r_q-1$ by removing $b-v$ $(\le n-1)$ points and q is any point of the plane.

2) L is a near pencil and q is any point of the near-pencil.

3) L is a simple extension of type 2 of a generalized projective plane and q is the special point.

4) L is an affine plane of order n with one point at infinity and q is not the point at infinity.

5) L is an s-fold inflated projective plane of order n for some integers s and n with $1 \le s < n$. If $D_1,\dots,D_s$ are its main subspaces then q is a point of every subspace D_j. Furthermore, if b_j is the number of lines of D_j, v_j its number of points, and r_j the degree of q in D_j, and if d is the deficiency of L, then $b_j \le v_j+r_j-2$ for all j, and

$$2+d+(s-1)(n+1)+\sum_{j=1}^{s}(b_j-v_j-r_j) \le 0.$$

Conversely every linear space satisfying 1), 2), 3), 4), or 5) is a linear space satisfying $b \le v+r_q-2$ for some point q.

An important part in the proof of the theorem plays the following lemma which follows easily from Lemma 2.2.

17.2 Lemma. Suppose a linear space has a point q such that $b \le v+r_q-2$ and such that $r_p \ge k_{pq}$ for every point $p \ne q$. Then q lies on a line L such that every point of L−{q} has degree k_L.■

The bound $b \le v+r_q-2$ in Lemma 17.2 is best possible. A counterexample for $b = v+r_q-1$ is a simple extension of type 1 of a projective plane where q is the special point of the extension.

Proof of Theorem 17.1. First we consider the case that $r_p \ge k_{pq}$ for every point $p \ne q$. Then Lemma 17.2 shows that there exists an integer n and a line L passing through q such that L and each point of L−{q} has degree n+1. If $n = 1$, then L and the point $\ne q$ of L have degree 2 so that L is a near-pencil. We may therefore assume that $n \ge 2$.

Since L has two points of degree n+1, every line has degree at most n+1. Set $v = n^2+c$. Then $b \le n^2+c+r_q-2$. Since L meets n^2+r_q lines, it follows that L is parallel to at most c−2 lines, which implies that $c \ge 2$. If p is any point of L−{q} then $v \ge n^2+c$ implies that p lies on at least c lines of degree n+1 with equality if and only if every line through p has degree n or n+1. We consider two cases.

Assume that L meets every other line and let $G \ne L$ be any line of degree n+1 which meets L in a point $\ne q$. Since L meets every other line, the Transfer Lemma implies that also G meets every other line. Hence q has degree $k_G = n+1$. Consequently every point of L has degree n+1, which implies that $b = n^2+n+1$. Corollary 3.5 shows that L can be extended to a projective plane of order n.

Now suppose that there is a line which is parallel to L. Since L is parallel to at most c−2 lines, it follows that $c \ge 3$. Because every point of degree n+1 is contained in $c \ge 2$ lines of degree n+1, every point has degree at least n+1. Consequently L satisfies the hypotheses of Theorem 16.1 and this theorem shows that L is an s-fold inflated projective plane of order n satisfying the conditions in 5).

From now on, we suppose that there exists a point $p \neq q$ such that the degree of p is strictly smaller than the degree of the line pq. Set $L = pq$, denote the degree of L by $n+1$ and the degree of p by $x+1$. Then $2 \leq x+1 \leq n$. We may assume that p is a point of minimal degree of $L-\{q\}$. Let q' be a point of minimal degree of the $n-1$ points of $L-\{q,p\}$ and denote its degree by $y+1$.

If $y = 1$ then $x = 1$ so that p and q' have degree 2. In this case, L is a near-pencil. From now on we assume that $y \geq 2$. Counting the number of lines which meet L, we see that

$$b \geq r_q+x+(n-1)y$$

with equality if and only if L meets every other line and if every point $\neq p,q$ of L has degree $y+1$. Counting the number of points from p, we see that

$$v \leq n+1+x \cdot y,$$

with equality if and only if every line $\neq L$ through p has degree $y+1$. We obtain

$$r_q+x+(n-1)y \leq b \leq v+r_q-2 \leq n+1+x \cdot y+r_q-2,$$

and hence

$$(n-1-x)(y-1) \leq 0.$$

In view of $x \leq n-1$ and $y \geq 2$, we obtain equality in all the above inequalities (except in the inequality $y \geq 2$). Hence $r_q+x+(n-1)y = b = v+r_q-2$, $v = n+1+x \cdot y$, and $x = n-1$. It follows that L meets every other line, that every point of $L-\{q,x\}$ has degree $y+1$, and that every line $\neq L$ through p has degree $y+1$.

Let s be any point $\neq p,q$ of L. Then s has degree $y+1$ and every line $\neq L$ through s has degree at most $r_p = x+1$. Since $v = n+1+x \cdot y$, we conclude that every line $\neq L$ through s has degree $x+1$.

Now let z be any point outside of L. Then the line zp has degree $y+1$ and the lines zs with $s \in L-\{q,p\}$ have degree $x+1$. Since L meets every other line, we obtain (count the number of points from z)

$$n+1+x \cdot y = v = k_{zq}+y+(n-1)x$$

so that

$$k_{zq} = 2+(x-1)(y+1-n).$$

Since the degree of zq can not exceed the degree x+1 of p and in view of $y \geq x = n-1$, we conclude that $y = x = n-1$ and $k_{zq} = 2$, or that $y = n$ and $k_{zq} = n$, or that $y = n+1$ and $k_{zq} = 2x$ and $x = 1$.

First consider the case $y = n$. Then every line other than L through q has degree n (since z was any point outside of L). Since $v = n+1+x \cdot y = n+1+(n-1)n$, it follows that q has degree n+1. We have shown now that every point other than p has degree n+1, that every line through p has degree $y+1 = n+1$, and that every other line has degree n. Consequently we obtain an affine plane of order n if we remove the point p. This shows that L is an affine plane of order n with the point p at infinity.

Now suppose that $y \neq n$. Then every line $\neq$ L through q has degree 2. Now we remove the point q and the 2-lines passing through q and obtain a linear space L' with $v' = v-1$ points and $b' = b+1-r_q$ lines. We have $b'-v' = b-(v+r_q-2) \leq 0$. Since every linear space has at least as much lines as points with equality if and only if it is a generalized projective plane, we conclude that $b = v+r_q-2$ and that L is a simple extension of type 2 of a generalized projective plane.■

For any non-incident point-line pair (p,L) of a linear space we denote by $\pi(p,L) = r_p-k_L$ the number of lines through p which miss L.

The Dowling-Wilson Conjecture (see Problem 2 in Erdös, Fowler, Sós and Wilson, 1985) says that every linear space has at least $\pi(p,L)$ more lines than points for each of its non-incident point-line pairs (p,L). As an appliction of Theorem 17.1, we shall prove this conjecture.

<u>17.3 Corollary</u> (Metsch, 1991c). Every linear space satisfies $b \geq v+\pi(p,L)$ for every non-incident point line pair (p,L).

This corollary will be proved together with Corollary 17.5. Corollary 17.3 gives rise to the following definition.

<u>17.4 Definition</u>. A linear space which satisfies $b = v+\pi(p,L)$ for one non-incident point line pair (p,L) will be called a *Dowling-Wilson* space.

<u>17.5 Corollary</u> (Metsch 1991c). Suppose that L is a Dowling-Wilson space and set $t = b-v$. Then one of the following cases occurs.

(1) L can be obtained from a projective plane of order $n \geq t+1$ by removing t points from a line.

(2) $t = 0$ and L is a near-pencil.

(3) L is the (3,t+2)-cross, i.e. $v = t+4$ and L has intersecting lines of degree 3 and t+2.

(4) L is an affine plane of order $n \geq t$ with a Dowling-Wilson space D at infinity. D has t−n more lines than points. If (p,L) is any non-incident point-line pair with $b = v+\pi(p,L)$ then p is a point of D and L is a line of D.

Conversely every linear space L satisfying (1), (2), (3), or (4) is a Dowling-Wilson space with t more lines than points.

<u>Proof</u> of Corollary 17.3 and 17.5. Suppose L is a linear space and q is a point and H a line of L with $q \notin H$ and $b \leq v+\pi(q,H)$. We have to show that $b = v+\pi(q,H)$ and that one of the cases in Corollary 17.5 occurs. We may assume that Corollary 17.3 is true for every linear space which has either less than v points or which has v points but less than b lines. Set $r = r_q$ and $k = k_D$. In view of $b \leq v+r-2$, one of the cases in Theorem 17.1 occurs.

First suppose that L can be obtained from a projective plane of n by removing $t := b-v$ points. Then $k_H = r-\pi(q,H) \leq r-(b-v) = n+1-t$. It follows that the t points have been removed from H and that $b = v+\pi(q,H)$.

The next case in Theorem 17.1 is that L is a near-pencil, which is clearly possible.

Next suppose that L is an affine plane of order n with a point at infinity. Then $b-v = n-1$. In view of $\pi(x,X) \leq 1$ for every non-incident point-line pair (x,X) in an affine plane with a point at infinity, we conclude that $n = 2$ and $t = 1$. Therefore $v = 5$ points and L is the (3,3)-cross.

A simple extension of type 2 of a projective plane of order n with special point x has r_x-2 more lines than points and every line of size two passes through the special point. This shows that a simple extension of type 2 of a projective plane satisfies $b > v+\pi(p,L)$ for every non-incident point-line pair. Therefore L is not a simple extension of type 2 of a projective plane. However, if $t := \pi(q,H)$ then

L may be the (3,t+2)-cross, which is a simple extension of type 2 of a near-pencil.

The last case in Theorem 16.1 is that L is an s-fold inflated projective plane for some integer s with $1 \le s \le n-1$ and that q is a point of all its main subspaces. Denote by d the defiency and by n the order of L (see the beginning of section 16 for the terminology). Furthermore let D be one of the main subspaces of L and let L be one of the n+1-s lines of L through q which are not lines of one of the main subspaces. Then L intersects each subspace only in q.

We denote by b_D the number of lines of D, by v_D its number of points, and by r_D the degree of q in D. The line L_0 meets n^2+r lines in L. Since every main subspace has at least one line which is parallel to L in L and because b_D-r_D is the number of lines of D which do not pass through q, we have

$b \ge n^2+r+b_D-r_D$ with equality if and only if s = 1. Furthermore

$v \le n^2+v_D$ and equality implies that d = 0.

Let k_D the degree of a shortest line H_D of D not passing through q. Because D has less than v points, Corollary 17.3 holds for D. Thus

$$b_D \ge v_D+\pi_D(p,H_D) = v_D+r_D-k_D.$$

We obtain a new linear space L' from L if we remove all the lines of D and replace it by a new line which contains all the points of D. Then L' has $v' := v$ points. Denote by b' the number of lines of L', by r' the degree of q in L', and by k' the degree of a shortest line H' of L' not passing through q. Since L' has v points and less than b lines, Corollary 17.3 holds for D. Hence

$$b' \ge v'+\pi'(q',H') = v'+r'-k'.$$

By hypothesis, we have

$$b \le v+\pi(q,H) = v+r-k.$$

From our definitions it follows that $b' = b+1-b_D$ and $r' = r+1-r_D$. Furthermore $k' \le k$ or $k_D \le k$. Using the above equations we obtain

$$v+r-k \geq b = b'+b_D-1 \geq v'+r'-k'+v_D+r_D-k_D-1 = v'-k'+v_D-k_D+r.$$

It follows that

$$k' \geq k+v_D-k_D > k.$$

Consequently $k \geq k_D$. In view of

$$n^2+v_D+r-k \geq v+r-k \geq b \geq n^2+r+b_D-r_D \geq n^2+r+v_D-k_D.$$

we obtain $k = k_D$ and $b_D-r_D = v_D-k_D$. Furthermore, $b = n^2+r+b_D-r_D$, which implies that $s = 1$, and $v = n^2+v_D$, which implies that $d = 0$. By definition, L is therefore an inflated affine plane of order n. In view of $b_D-r_D = v_D-k_D$, D is also a Dowling-Wilson space. Because of $k' > k$ we see furthermore that H is a line of D. This completes the proof of Corollary 17.3 and Corollary 17.5.■

18. Uniqueness of embeddings

Consider the linear space L with 8 points and two parallel lines H and L of degree 4. Then L can be extended to a projective plane P of order 4. Furthermore, P is uniquely determined, since there is only one projective plane of order 4. However, there is more than one possibility to extend L to P. If p_j are the points of G and q_j are the points of H, then we may extend L to P such that the lines p_1q_1, p_2q_2, p_3q_3, p_4q_4 are concurrent in P, or such that the lines $p_1q_1, p_2q_2, p_3q_4, p_4q_3$ are concurrent in P. But both can never be satisfied at the same time.

We say that a linear space L can be *uniquely embedded* into a projective plane of order n if it can be embedded and if for any two projective planes P_1 and P_2 of order n in which L is embedded there exists an ismorphism from P_1 onto P_2 which fixes every point and line of L.

We want to show that every embedding of a linear space into a projective plane of order n is unique if it has at least n^2-n+2 points. We begin with the following lemma, which is similar to Lemma 7.1.

18.1 Lemma. Let P be a projective plane and denote by n its order. Suppose that $p \cup q$ is a partition of the set of points and Π is a set lines such that every point of p lies on a unique line of Π. If $|\Pi| \le n+1$ and $|p| \ge n^2-n+1$, then one of the following cases occurs.

a) The lines of Π pass through a common point.

b) $|\Pi| = n+1$, $|p| = n^2-n+1$, and there are two lines G and H such that q consists of the $2n$ points $\neq G\cap H$ of $G\cup H$. Furthermore, it exists a point x on G such that Π consists of the line H and the n lines $\neq$ G passing through x.

Proof. If Π contains a line L which has only lines in p, then $\Pi = \{L\}$, and if Π contains a line N with a unique point q in q, then each line of Π passes through q. We may therefore assume that every line of Π has at most $n-1$ points in p. Since the lines of Π cover the $|p| \ge n^2-n+1$ points of p, it follows that $|\Pi| = n+1$ and that Π contains two lines L_1 and L_2 with $n-1$ points in p. Let q be the point of intersection of L_1 and L_2, which is in q, and denote by q_j the second point of $L_j \cap q$. Then every line of Π intersects L_j in q or q_j. Hence, $H := q_1q_2$ is the only line not passing through q which could be in Π. Thus, if $H \notin \Pi$, then we are in case a). If $H \in \mathbb{T}$, then the n lines of $\mathbb{T}-\{H\}$ pass through q, so q lies on a unique line G which is not in $\mathbb{T}$. The n−1 points $\neq q$, $G \cap H$ of G are in q, since they do not lie on any line of $\mathbb{T}$ and q is in q, since it lies on n lines of $\mathbb{T}$. Furthermore, the n points $\neq G \cap H$ of H are in q, since they lie on two lines of $\mathbb{T}$. In view of $|q| \leq 2n$, it follows that $q = (G \cup H)-(G \cap H)$. Hence, b) is fulfilled.■

<u>18.2 Theorem</u>. Suppose that L is a linear space with at least n^2-n+1 points which is embedded in a projective plane of order n. Then the embedding is uniquely determined.

<u>Proof</u>. Suppose that $L = (p,L)$ is embedded in two projective planes $P_1 = (p_1,L_1)$ and $P_2 = (p_2,L_2)$. We have to show that there is an isomorphism from P_1 onto P_2 which fixes the points and lines of L. Notice that p is contained in p_1 and p_2. We consider two cases.

Case 1. $p_2 - p \neq (G_2 \cup H_2)-(G_2 \cap H_2)$ for all lines G_2 and H_2 of P_2.
If p is a point of L, then the number of lines of P_j which meet p only in p does not depend on j. Also the number of lines of P_j which have no point in p does not depend on j. It follows that there is a bijection f from L_1 onto L_2 such that $f(L) = L$ for all $L \in L$ and such that L and f(L) contain the same points of p for all $L \in L_1$. We shall extend f to an isomorphism from P_1 onto P_2. If p is a point of p, then set $f(p) := p$. If p be a point of $p_1 - p$, then we define f(p) as follows. Denote by $\mathbb{T}_p$ the set of the n+1 lines of P_1 through p. Then each point of p lies on a unique line of $\mathbb{T}_p$. Hence each point of p lies on a unique line of $f(\mathbb{T}_q)$. Lemma 18.1 shows that the lines of $f(\mathbb{T})$ are concurrent in P_2. It follows that f becomes an isomorphism from P_1 onto P_2, if we let f(p) denote the point of intersection of the lines of $f(\mathbb{T}_p)$ for all $p \in p_1 - p$.

Case 2. There are lines G_2 and H_2 with $p_2 - p = (G_2 \cup H_2)-(G_2 \cap H_2)$.
Then there are also lines G_1 and H_1 with $p_1 - p = (G_1 \cup H_1)-(G_1 \cap H_1)$. Set $x = G_j \cap H_j \in p$. Let q be a point of $G_1-\{x\}$, and denote by $\mathbb{T}$ the set of lines $\neq G_1$ through q. Then each of point of L lines on a unique line of $\mathbb{T}$. In view of $\mathbb{T} \subseteq L \subseteq L_2$, Lemma 18.1 shows that the lines of $\mathbb{T} \cup \{G_2\}$ or that the lines of $\mathbb{T} \cup \{H_2\}$ intersect in P_2 in a point. Hence, the lines of $\mathbb{T}$ intersect in P_2 in a point $q' \in H_2 \cup G_2 - \{x\}$. Similarly, the lines $\neq H_1$ passing through a point q of $H_1-\{x\}$ meet in P_2 in a point $q' \neq x$ of

$H_2 \cup G_2$. This gives a correspondence of the points $q \in G_1 \cup H_1 - \{x\}$ and the points $q' \in G_2 \cup H_2 - \{x\}$. Furthermore, q_1 and q_2 lie on the same line of $\{G_1, H_1\}$ if and only if q'_1 and q'_2 are on the same line of $\{G_2, H_2\}$ (only in this case the corresponding sets $\mathbb{T}_1$ and $\mathbb{T}_2$ have no line in common). We may therefore assume that the points of G_1 (or H_1) correspond to the points of G_2 (or H_2). It is now obvious that the identity map on L extends to an isomorphism f from P_1 onto P_2 with $f(q) = q'$ for all $q \in G_1 \cup H_1 - \{x\}$, $f(G_1) = G_2$ and $f(H_1) = H_2$.■

Remark. Every line of a linear space has at least two points. This property of linear spaces has not been used in the preceding proof. Thus, Theorem 18.2 rests also true for 'linear spaces' with lines of degree 1. We shall use this fact in the proof of Corollary 18.3.

We have also embedded linear spaces into the closed complement of a Baer subplane in a projective plane. We shall show that these embedding are also 'uniquely' determined.

18.3 Corollary. Suppose L is a linear space with $v \geq n^2-n+2$ and a point q which has degree at least n+2. If L can be embedded into the closed complement D of a Baer subplane B in a projective plane P of order $n = m^2$, then the embedding is unique.

Proof. Let D' be any complement of Baer subplane B' in a projective plane P' of order $n = m^2$ such that L can be embedded into D'. We have to show that there is an isomorphism from D onto D' which is the identity on L.

L has n^2+n+1 lines (this follows from $v \geq n^2-n+2$) and the point q has degree n+1+m. The lines through q are the lines of B. $L-\{q\}$ is a linear space with at least n^2-n+1 points, which is embedded in the complement of B in P and in the complement of B' of P'. By Theorem 18.2, there is an isomorphism f from P onto P' fixing every point and line of $L-\{q\}$ (by the preceding remark, it does not disturb that $L-\{q\}$ might have a line of degree 1).

Let x be any point of the subplane B of P, and let H and L be two of the m+1 lines of B through x. Then $f(x) = f(H \cap L) = f(H) \cap f(L)$. Since H and L are lines of $L-\{q\}$, which are fixed by f, $f(H) \cap f(L)$ is the point of intersection of H and L in

P'. Because H and L are also lines of the Baer-subplane B' (this follows from the definition of the closed complement of B' in P' and the fact that H and L pass through q), it follows that f(x) is a point of B'. Hence, f sends points of B to points of B'.

This implies that the restriction f^* of f on P−B is an isomorphism from P−B onto P'−B fixing L−{q}. If we set $f^*(q) := q$, then f becomes an isomorphism from D onto D', which fixes L.■

We have also studied linear spaces with $n^2-n+2 \le v \le n^2+n+1 < b$. The only result we obtained is that some of them are inflated affine planes of order n so that there are no embeddings which should be proved to be unique. However, notice the following. Suppose P is a projective plane of order n, p_1 and p_2 are two points of a line L of P, and P has no automorphism which fixes L and sends p_1 to p_2. Let A be the affine plane P−L and let D be the near-pencil on n+1 points. D has a unique point q of degree n. Let L_j, j = 1,2 be linear space consisting of A with D at infinity such that q corresponds to p_j. Then L_1 and L_2 are not isomorphic (Assume to the contrary that there is an isomorphism α from L_1 onto L_2. Since p_j is the only point of degree 2n of L_j, we have $\alpha(p_1)= \alpha(p_2)$. If we set $\alpha(L) = L$, then the restriction of α to P is an automorphism of P. But we assumed that there is no such isomorphism, a contradiction). This example shows that an inflated affine plane is not uniquely determined by the affine plane and the linear space at infinity.

REFERENCES

BAKER, C.A./BLOKHUIS, A./BROUWER, A.E./WILBRINK, H.A.
1990 Characterization Theorems for failed projective and affine planes. Coding Theory and Design Theory Part II (Ed. D. Ray-Chaudhuri), pp. 42-53, Springer, New York, 1990.

BASTERFIELD, J.G./KELLY, L.M.
1968 A characterization of sets of n points which determine n hyperplanes. Proc. Camb. Phil. Soc. 64, 585-588.

BATTEN, L.M./SANE, S.S.
1985 A characterization of the complement of the union of two disjoint Baer subplanes. Arch. Math. 44, 569-576.

BETH, Th./JUNGNICKEL, D./LENZ, H.
1985 Design Theory. Bibliographisches Institut, Mannheim-Wien-Zürich, 1985.

BEUTELSPACHER, A.
1982 Einführung in die endliche Geometrie I: Blockpläne. Bibliographisches Institut Mannheim-Wien-Zürich, 1982.
1983 Einführung in die endliche Geometrie II: Projektive Räume. Bibliographisches Institut Mannheim-Wien-Zürich, 1983.

BEUTELSPACHER, A./METSCH, K.
1986 Embedding finite linear spaces in projective planes. Ann. Discrete Math. 30, 39-50.
1987 Embedding finite linear spaces in projective planes, II. Discrete Math. 66, 219-230.

BLOKHUIS, A./SCHMIDT, R.J.M./WILBRINK, H.A.
1988 On the number of lines in a linear space on p^2+p+1 points. Proceedings of Cambinatorics 88 held in Ravello, 23-28 May 1988, to appear.

BOSE, R.C.
1963 Strongly regular graphs, partial geometries and partially balanced designs. Pacific J. Math. 13, 389-419.

BOSE, R.C./SHRIKHANDE, S.S.
1973 Embedding the complement of an oval in a projective plane of even order. Discrete Math. 6, 305-312.

BOUTEN, M./DE WITTE, P.
1965 A new proof of an inequality of Szekeres, de Bruijn and Erdős. Bull. Soc. Math. Belg. XVII, 475-483.

BRIDGES, W.G.
1972 Near 1-designs. J. Comb. Theory (Series A) 13, 116-126.

BRUCK, R.H./RYSER, H.J.
1949 The nonexistence of certain finite projective planes. Math. Z. 119, 273-275.

BRUCK, R.H.
1963 Finite nets II: Uniqueness and imbedding. Pacific J. Math. 13, 421-457.

BRUEN, A.
1971 Blocking sets in finite projective planes. SIAM, J. Appl. Math. 21, 380-392.
1973 The number of lines determined by n^2 points. J. Comb. Theory (A) 15, 225-241.

BRUIJN, N.G. DE/ERDÖS, P.
1948 On a combinatorial problem. Nederl. Akad. Wetensch. Indag. Math. 10, 1277-1279.

CHOWLA, S./RYSER, H.J.
1950 Combinatorial problems. Canad. J. Math. 2, 93-99.

DE VITO, P./LO RE, P.M.
1988a On a class of linear spaces. Proceedings of Combinatorics 88 held in Ravello, 23-28 May 1988, to appear.
1988b Spazi lineari su v punti in cui ogni retta di lunghezza massima é intersecata da al più v+1 altre rette. Rendiconti di Matem. Serie VII, vol. 8, 455-466.

DE VITO, P./LO RE, P.M./METSCH, K.
1991 Linear spaces in which every long line intersects v+2 other lines. Arch. Math., to appear.

DOW, S.
1982 Extending partial projective planes. Congressus Numerantium 35, 245-251.
1983 An improved bound for extending partial projective planes. Discrete Math. 45, 199-207.

DRAKE, D.A.
1985 Embedding pseudo-complements of disjoint pairs of Baer subplanes. Algebras Groups Geom. 2, 278-294.

ERDÖS, P./MULLIN, R.C./SOS, V.T/STINSON, D.R.
1983 Finite linear spaces and projective planes. Discrete Math. 47, 49-62.

ERDÖS, P./FOWLER, J.C./SOS, V.T./WILSON, R.M.
1985 On 2-Designs. J. Comb. Theory (A) 38, 131-142.

FOWLER, J.C.
1984 A short proof of Totten's Classification of restricted linear spaces. Geom. Dedicata 15, 413-422.

FÜREDI, Z.
1990 Quadrilateral free graphs with maximum possible number of edges, to appear.

HUGHES, D.R./PIPER, F.C.
1973 Projective Planes. Springer, Berlin-Heidelberg-New York, 1973.

HANANI, H.
1951 On the number of straight lines determined by n points. Riveon Lematematika 5, 10-11.

HANANI, H.
1954/55 On the number of lines and planes determined by d points. Scientific Publications, Technion (Isreal Institute of Technology, Haifa) 6, 58-63.

LI-CHIEN, C.
1960 Association schemes of partially balanced designs with parameters v=28, n_1=12, n_2=15 and p_{11}=4. Science Record (Peking), Math. New Ser. 4, 12-18.

MAJUMDAR, K.N.
1953 On some theorems in combinatorics relating to incomplete block designs. Ann. Math. Stat. 24, 377-389.

McCARTHY, D.
1976 Transversals in Latin squares of order 6 and (7,1)-designs. Ars Combinatoria 1, 3-100.

McCARTHY, D./Mullin, R.C.
1976 The Non-Existence of (7,1)-designs with $v = 31$ and $b \leq 43$. Proceedings of the Fifth Manitoba Conference on Numerical Mathematics (Univ. Manitoba, Winnipeg, Man., 1975), pp. 479-495. Congressus Numerantium, No. XVI, Utilitas Math. Publ., Winnipeg, Man., 1976.

McCARTHY, D./MULLIN, R.C./SCHELLENBERG, P.J./STANTON, R.G./ VANSTONE, S.A.
1976a The Non-Existence of (7,1)-designs with $v = 31$ and $b \geq 50$', Proceedings of the Fifth Manitoba Conference on Numerical Mathematics (Univ. Manitoba, Winnipeg, Man., 1975), pp. 497-532. Congressus Numerantium, No. XVI, Utilitas Math. Publ., Winnipeg, Man., 1976.
1976b On approximations to a Projective plane of order 6. Ars Combinatoria 2, 111-168.

McCARTHY, D./VANSTONE, S.A.
1977 Embedding (r,1)-Designs in finite projective planes. Discrete Math. 19, 67-76.

MELONE, N.
1987 Un teoremna di struttura per gli spazi lineari. Pubbl. del Dip. di Matem. e appl. "R. Caccioppoli" Napoli 58.

MENDELSOHN, E./ASSAF, A.
1987 Spectrum of Imbrical Designs. Annals of Discete Math. 34, 363-370.

METSCH, K.
1988a An improved bound for the embedding of linear spaces into projective planes', Geom. Ded. 26, 333-340.
1988b An optimal bound for embedding linear spaces into projective planes. Discrete Math. 70, 53-70.
1988c Extending the complement of any number of mutually disjoint Baer-subplanes to a projective plane. Proceedings of Combinatorics 88 held in Ravello, 23-28 May 1988, to appear.
1990a On the non-existence of certain (7,1)-designs. Proceedings of the 2nd international Catania Combinatorial Conference on Graps, Designs and Combinatorial Geometries held in Catania-Santa Tecla, Setp. 4-9 1989, Journal of Combinatorics, Information and System Science 15, 70-84.
1990b Linear spaces in which every line of maximal degree meets only few lines. Proceedings of Combinatorics 90 held in Gaeta, 20-27 May 1990, to appear.
1991a Linear spaces with few lines. Discrete Math., to appear.
1991b Linear spaces with n^2+n+2 lines, European J. Comb., to appear.
1991c The Dowling-Wilson conjecture. Bull. Soc. Math. Belg., to appear.
1991d Improvement of Bruck's Completion Theorem. Designs, Codes and Cryptography 1, to appear.

MULLIN, R.C./VANSTONE, S.A.
1975 On regular pairwise balanced designs of order 6 and index 1. Utilitas Math. 8, 349-369.
1976a On the non-existence of a certain design. Utilitas Math. 9, 193-207.
1976b A generalization of a theorem of Totten. J. Austral. Math. Soc. 22, 494-500.

RIGBY, J.F.
1965 Affine subplanes of projective planes. Canad. J. Math. 17, 977-1014.

SCHÜTZENBERGER, M.P.
1949 A non-existence criterion for an infinite family of symmetrical block designs. Ann. Eugenics 14, 286-287.

SHRIKHANDE, S.S.
1969 The uniqueness of the L_2 association scheme. Ann. Math. Stat. 30, 781-798.

STANTON, R.G./KALBFLEISCH, J.G.
1972 The $\lambda-\mu$ problem: $\lambda=1$ and $\mu=3$. Proc. Second Chapel Hill Conf. on Combinatorics, Chapell Hill (1972), 451-462.

STINSON, D.R.
1983 The non-existence of certain finite linear spaces. Geom. Ded. 13, 429-434.

TARRY, G.
1900 Le probléme de 36 officiers. C.R. Assoc. Franc. Avanc. Sci. nat. 1, 122-123.

THAS, J.A./DE CLERCK, F.
1957 Some applications of the fundamental characterization theorem of R.C. Bose to partial geometries. Lincei - Rend. Sc. fis. mat. e nat. 59, 86-90.

TOTTEN, J.
1975 Basic properties of restricted linear spaces. Discrete Math. 13, 67-74.
1976a On the degree of points and lines in restricted linear spaces. Discrete Math. 14, 391-394.
1976b Parallelism in restricted linear spaces. Discrete Math. 14, 395-398.
1976c Classification of restricted linear spaces. Can. J. Math. 28, 321-333.
1976d Finite linear spaces with three more lines than points. Simon Stevin 47, 35-47.
1976e Embedding the complement of two lines in a finite projective plane. J. Austral. Math. Soc. (Series A) 22, 27-34.

VANSTONE, S.A.
1973 The Extendability of (r,1)-Designs. Proc. Third Manitoba Conf. on Numerical Math., Winnipeg 1973, pp. 409-418.

VARGA, L.E.
1985 A note on the structure of Pairwise Balanced Designs. J. Comb. Th. (A) 40, 435-438.

WITTE, P. DE
1975a Restricted linear spaces with a square number of points. Simon Stevin 48, 107-120.
1975b Combinatorial properties of Finite Linear Spaces II. Bull. Soc. Math. Belg. 27, 115-155.
1976 Finite linear spaces with two more lines than points. J. reine angew. Math. 288, 66-73.
1977a The Exceptional case in a Theorem of Bose and Shrikhande. J. Austral. Math. Soc. (Series A) 24, 64-78.
1977b On the embedding of linear spaces in projective planes of order n. Trans. Amer. Math. Soc., unpublished.

INDEX

Lecture Notes in Mathematics

For information about Vols. 1–1296 please contact your bookseller or Springer-Verlag

Vol. 1297: Zhu Y.-l., Guo B.-y. (Eds.), Numerical Methods for Partial Differential Equations. Proceedings. XI, 244 pages. 1987.

Vol. 1298: J. Aguadé, R. Kane (Eds.), Algebraic Topology, Barcelona 1986. Proceedings. X, 255 pages. 1987.

Vol. 1299: S. Watanabe, Yu.V. Prokhorov (Eds.), Probability Theory and Mathematical Statistics. Proceedings, 1986. VIII, 589 pages. 1988.

Vol. 1300: G.B. Seligman, Constructions of Lie Algebras and their Modules. VI, 190 pages. 1988.

Vol. 1301: N. Schappacher, Periods of Hecke Characters. XV, 160 pages. 1988.

Vol. 1302: M. Cwikel, J. Peetre, Y. Sagher, H. Wallin (Eds.), Function Spaces and Applications. Proceedings, 1986. VI, 445 pages. 1988.

Vol. 1303: L. Accardi, W. von Waldenfels (Eds.), Quantum Probability and Applications III. Proceedings, 1987. VI, 373 pages. 1988.

Vol. 1304: F.Q. Gouvêa, Arithmetic of p-adic Modular Forms. VIII, 121 pages. 1988.

Vol. 1305: D.S. Lubinsky, E.B. Saff, Strong Asymptotics for Extremal Polynomials Associated with Weights on R. VII, 153 pages. 1988.

Vol. 1306: S.S. Chern (Ed.), Partial Differential Equations. Proceedings, 1986. VI, 294 pages. 1988.

Vol. 1307: T. Murai, A Real Variable Method for the Cauchy Transform, and Analytic Capacity. VIII, 133 pages. 1988.

Vol. 1308: P. Imkeller, Two-Parameter Martingales and Their Quadratic Variation. IV, 177 pages. 1988.

Vol. 1309: B. Fiedler, Global Bifurcation of Periodic Solutions with Symmetry. VIII, 144 pages. 1988.

Vol. 1310: O.A. Laudal, G. Pfister: Local Moduli and Singularities. V, 117 pages. 1988.

Vol. 1311: A. Holme, R. Speiser (Eds.), Algebraic Geometry, Sundance 1986. Proceedings. VI, 320 pages. 1988.

Vol. 1312: N.A. Shirokov, Analytic Functions Smooth up to the Boundary. III, 213 pages. 1988.

Vol. 1313: F. Colonius. Optimal Periodic Control. VI, 177 pages. 1988.

Vol. 1314: A. Futaki. Kähler-Einstein Metrics and Integral Invariants. IV, 140 pages. 1988.

Vol. 1315: R.A. McCoy, I. Ntantu, Topological Properties of Spaces of Continuous, Functions. IV, 124 pages. 1988.

Vol. 1316: H. Korezlioglu, A.S. Ustunel (Eds.), Stochastic Analysis and Related Topics. Proceedings, 1986. V, 371 pages. 1988.

Vol. 1317: J. Lindenstrauss, V.D. Milman (Eds.), Geometric Aspects of Functional Analysis. Seminar, 1986–87. VII, 289 pages. 1988.

Vol. 1318: Y. Felix (Ed.), Algebraic Topology – Rational Homotopy. Proceedings, 1986. VIII, 245 pages. 1988.

Vol. 1319: M. Vuorinen, Conformal Geometry and Quasiregular Mappings. XIX, 209 pages. 1988.

Vol. 1320: H. Jürgensen, G. Lallement, H.J. Weinert (Eds.), Semigroups, Theory and Applications. Proceedings, 1986. X, 416 pages. 1988.

Vol. 1321: J. Azéma, P.A. Meyer, M. Yor (Eds.), Séminaire de Probabilités XXII. Proceedings. IV, 600 pages. 1988.

Vol. 1322: M. Métivier, S. Watanabe (Eds.), Stochastic Analysis. Proceedings, 1987. VII, 197 pages. 1988.

Vol. 1323: D.R. Anderson, H.J. Munkholm, Boundedly Controlled Topology. XII, 309 pages. 1988.

Vol. 1324: F. Cardoso, D.G. de Figueiredo, R. Iório, O. Lopes (Eds.), Partial Differential Equations. Proceedings, 1986. VIII, 433 pages. 1988.

Vol. 1325: A. Truman, I.M. Davies (Eds.), Stochastic Mechanics and Stochastic Processes. Proceedings, 1986. V, 220 pages. 1988.

Vol. 1326: P.S. Landweber (Ed.), Elliptic Curves and Modular Forms in Algebraic Topology. Proceedings, 1986. V, 224 pages. 1988.

Vol. 1327: W. Bruns, U. Vetter, Determinantal Rings. VII, 236 pages. 1988.

Vol. 1328: J.L. Bueso, P. Jara, B. Torrecillas (Eds.), Ring Theory. Proceedings, 1986. IX, 331 pages. 1988.

Vol. 1329: M. Alfaro, J.S. Dehesa, F.J. Marcellan, J.L. Rubio de Francia, J. Vinuesa (Eds.): Orthogonal Polynomials and their Applications. Proceedings, 1986. XV, 334 pages. 1988.

Vol. 1330: A. Ambrosetti, F. Gori, R. Lucchetti (Eds.), Mathematical Economics. Montecatini Terme 1986. Seminar. VII, 137 pages. 1988.

Vol. 1331: R. Bamón, R. Labarca, J. Palis Jr. (Eds.), Dynamical Systems, Valparaiso 1986. Proceedings. VI, 250 pages. 1988.

Vol. 1332: E. Odell, H. Rosenthal (Eds.), Functional Analysis. Proceedings. 1986–87. V, 202 pages. 1988.

Vol. 1333: A.S. Kechris, D.A. Martin, J.R. Steel (Eds.), Cabal Seminar 81–85. Proceedings, 1981–85. V, 224 pages. 1988.

Vol. 1334: Yu.G. Borisovich, Yu.E. Gliklikh (Eds.), Global Analysis – Studies and Applications III. V, 331 pages. 1988.

Vol. 1335: F. Guillén, V. Navarro Aznar, P. Pascual-Gainza, F. Puerta, Hyperrésolutions cubiques et descente cohomologique. XII, 192 pages. 1988.

Vol. 1336: B. Helffer, Semi-Classical Analysis for the Schrödinger Operator and Applications. V, 107 pages. 1988.

Vol. 1337: E. Sernesi (Ed.), Theory of Moduli. Seminar, 1985. VIII, 232 pages. 1988.

Vol. 1338: A.B. Mingarelli, S.G. Halvorsen. Non-Oscillation Domains of Differential Equations with Two Parameters. XI, 109 pages. 1988.

Vol. 1339: T. Sunada (Ed.), Geometry and Analysis of Manifolds. Proceedings, 1987. IX, 277 pages. 1988.

Vol. 1340: S. Hildebrandt, D.S. Kinderlehrer, M. Miranda (Eds.), Calculus of Variations and Partial Differential Equations. Proceedings, 1986. IX, 301 pages. 1988.

Vol. 1341: M. Dauge, Elliptic Boundary Value Problems on Corner Domains. VIII, 259 pages. 1988.

Vol. 1342: J.C. Alexander (Ed.), Dynamical Systems. Proceedings, 1986–87. VIII, 726 pages. 1988.

Vol. 1343: H. Ulrich, Fixed Point Theory of Parametrized Equivariant Maps. VII, 147 pages. 1988.

Vol. 1344: J. Král, J. Lukes, J. Netuka, J. Vesely´ (Eds.), Potential Theory – Surveys and Problems. Proceedings, 1987. VIII, 271 pages. 1988.

Vol. 1345: X. Gomez-Mont, J. Seade, A. Verjovski (Eds.), Holomorphic Dynamics. Proceedings, 1986. VII. 321 pages. 1988.

Vol. 1346: O.Ya. Viro (Ed.), Topology and Geometry – Rohlin Seminar. XI, 581 pages. 1988.

Vol. 1347: C. Preston, Iterates of Piecewise Monotone Mappings on an Interval. V, 166 pages. 1988.

Vol. 1348: F. Borceux (Ed.), Categorical Algebra and its Applications. Proceedings, 1987. VIII, 375 pages. 1988.

Vol. 1349: E. Novak, Deterministic and Stochastic Error Bounds in Numerical Analysis. V, 113 pages. 1988.

Vol. 1350: U. Koschorke (Ed.), Differential Topology Proceedings, 1987, VI, 269 pages. 1988.

Vol. 1351: I. Laine, S. Rickman, T. Sorvali (Eds.), Complex Analysis, Joensuu 1987. Proceedings. XV, 378 pages. 1988.

Vol. 1352: L.L. Avramov, K.B. Tchakerian (Eds.), Algebra – Some Current Trends. Proceedings. 1986. IX, 240 Seiten. 1988.

Vol. 1353: R.S. Palais, Ch.-l. Teng, Critical Point Theory and Submanifold Geometry. X, 272 pages. 1988.

Vol. 1354: A. Gómez, F. Guerra, M.A. Jiménez, G. López (Eds.), Approximation and Optimization. Proceedings, 1987. VI, 280 pages. 1988.

Vol. 1355: J. Bokowski, B. Sturmfels, Computational Synthetic Geometry. V, 168 pages. 1989.

Vol. 1356: H. Volkmer, Multiparameter Eigenvalue Problems and Expansion Theorems. VI, 157 pages. 1988.

Vol. 1357: S. Hildebrandt, R. Leis (Eds.), Partial Differential Equations and Calculus of Variations. VI, 423 pages. 1988.

Vol. 1358: D. Mumford, The Red Book of Varieties and Schemes. V, 309 pages. 1988.

Vol. 1359: P. Eymard, J.-P. Pier (Eds.) Harmonic Analysis. Proceedings, 1987. VIII, 287 pages. 1988.

Vol. 1360: G. Anderson, C. Greengard (Eds.), Vortex Methods. Proceedings, 1987. V, 141 pages. 1988.

Vol. 1361: T. tom Dieck (Ed.), Algebraic Topology and Transformation Groups. Proceedings. 1987. VI, 298 pages. 1988.

Vol. 1362: P. Diaconis, D. Elworthy, H. Föllmer, E. Nelson, G.C. Papanicolaou, S.R.S. Varadhan. École d' Été de Probabilités de Saint-Flour XV–XVII. 1985–87 Editor: P.L. Hennequin. V, 459 pages. 1988.

Vol. 1363: P.G. Casazza, T.J. Shura, Tsirelson's Space. VIII, 204 pages. 1988.

Vol. 1364: R.R. Phelps, Convex Functions, Monotone Operators and Differentiability. IX, 115 pages. 1989.

Vol. 1365: M. Giaquinta (Ed.), Topics in Calculus of Variations. Seminar, 1987. X, 196 pages. 1989.

Vol. 1366: N. Levitt, Grassmannians and Gauss Maps in PL-Topology. V, 203 pages. 1989.

Vol. 1367: M. Knebusch, Weakly Semialgebraic Spaces. XX, 376 pages. 1989.

Vol. 1368: R. Hübl, Traces of Differential Forms and Hochschild Homology. III, 111 pages. 1989.

Vol. 1369: B. Jiang, Ch.-K. Peng, Z. Hou (Eds.), Differential Geometry and Topology. Proceedings, 1986–87. VI, 366 pages. 1989.

Vol. 1370: G. Carlsson, R.L. Cohen, H.R. Miller, D.C. Ravenel (Eds.), Algebraic Topology. Proceedings, 1986. IX, 456 pages. 1989.

Vol. 1371: S. Glaz, Commutative Coherent Rings. XI, 347 pages. 1989.

Vol. 1372: J. Azéma, P.A. Meyer, M. Yor (Eds.), Séminaire de Probabilités XXIII. Proceedings. IV, 583 pages. 1989.

Vol. 1373: G. Benkart, J.M. Osborn (Eds.), Lie Algebras. Madison 1987. Proceedings. V, 145 pages. 1989.

Vol. 1374: R.C. Kirby, The Topology of 4-Manifolds. VI, 108 pages. 1989.

Vol. 1375: K. Kawakubo (Ed.), Transformation Groups. Proceedings, 1987. VIII, 394 pages, 1989.

Vol. 1376: J. Lindenstrauss, V.D. Milman (Eds.), Geometric Aspects of Functional Analysis. Seminar (GAFA) 1987–88. VII, 288 pages. 1989.

Vol. 1377: J.F. Pierce, Singularity Theory, Rod Theory, and Symmetry-Breaking Loads. IV, 177 pages. 1989.

Vol. 1378: R.S. Rumely, Capacity Theory on Algebraic Curves. III, 437 pages. 1989.

Vol. 1379: H. Heyer (Ed.), Probability Measures on Groups IX. Proceedings, 1988. VIII, 437 pages. 1989.

Vol. 1380: H.P. Schlickewei, E. Wirsing (Eds.), Number Theory, Ulm 1987. Proceedings. V, 266 pages. 1989.

Vol. 1381: J.-O. Strömberg, A. Torchinsky, Weighted Hardy Spaces. V, 193 pages. 1989.

Vol. 1382: H. Reiter, Metaplectic Groups and Segal Algebras. XI, 128 pages. 1989.

Vol. 1383: D.V. Chudnovsky, G.V. Chudnovsky, H. Cohn, M.B. Nathanson (Eds.), Number Theory, New York 1985–88. Seminar. V, 256 pages. 1989.

Vol. 1384: J. Garcia-Cuerva (Ed.), Harmonic Analysis and Partial Differential Equations. Proceedings, 1987. VII, 213 pages. 1989.

Vol. 1385: A.M. Anile, Y. Choquet-Bruhat (Eds.), Relativistic Fluid Dynamics. Seminar, 1987. V, 308 pages. 1989.

Vol. 1386: A. Bellen, C.W. Gear, E. Russo (Eds.), Numerical Methods for Ordinary Differential Equations. Proceedings, 1987. VII, 136 pages. 1989.

Vol. 1387: M. Petkovi´c, Iterative Methods for Simultaneous Inclusion of Polynomial Zeros. X, 263 pages. 1989.

Vol. 1388: J. Shinoda, T.A. Slaman, T. Tugué (Eds.), Mathematical Logic and Applications. Proceedings, 1987. V, 223 pages. 1989.

Vol. 1000: Second Edition. H. Hopf, Differential Geometry in the Large. VII, 184 pages. 1989.

Vol. 1389: E. Ballico, C. Ciliberto (Eds.), Algebraic Curves and Projective Geometry. Proceedings, 1988. V, 288 pages. 1989.

Vol. 1390: G. Da Prato, L. Tubaro (Eds.), Stochastic Partial Differential Equations and Applications II. Proceedings, 1988. VI, 258 pages. 1989.

Vol. 1391: S. Cambanis, A. Weron (Eds.), Probability Theory on Vector Spaces IV. Proceedings, 1987. VIII, 424 pages. 1989.

Vol. 1392: R. Silhol, Real Algebraic Surfaces. X, 215 pages. 1989.

Vol. 1393: N. Bouleau, D. Feyel, F. Hirsch, G. Mokobodzki (Eds.), Séminaire de Théorie du Potentiel Paris, No. 9. Proceedings. VI, 265 pages. 1989.

Vol. 1394: T.L. Gill, W.W. Zachary (Eds.), Nonlinear Semigroups, Partial Differential Equations and Attractors. Proceedings, 1987. IX, 233 pages. 1989.

Vol. 1395: K. Alladi (Ed.), Number Theory, Madras 1987. Proceedings. VII, 234 pages. 1989.

Vol. 1396: L. Accardi, W. von Waldenfels (Eds.), Quantum Probability and Applications IV. Proceedings, 1987. VI, 355 pages. 1989.

Vol. 1397: P.R. Turner (Ed.), Numerical Analysis and Parallel Processing. Seminar, 1987. VI, 264 pages. 1989.

Vol. 1398: A.C. Kim, B.H. Neumann (Eds.), Groups – Korea 1988. Proceedings. V, 189 pages. 1989.

Vol. 1399: W.-P. Barth, H. Lange (Eds.), Arithmetic of Complex Manifolds. Proceedings, 1988. V, 171 pages. 1989.

Vol. 1400: U. Jannsen. Mixed Motives and Algebraic K-Theory. XIII, 246 pages. 1990.

Vol. 1401: J. Steprans, S. Watson (Eds.), Set Theory and its Applications. Proceedings, 1987. V, 227 pages. 1989.

Vol. 1402: C. Carasso, P. Charrier, B. Hanouzet, J.-L. Joly (Eds.), Nonlinear Hyperbolic Problems. Proceedings, 1988. V, 249 pages. 1989.

Vol. 1403: B. Simeone (Ed.), Combinatorial Optimization. Seminar, 1986. V, 314 pages. 1989.

Vol. 1404: M.-P. Malliavin (Ed.), Séminaire d'Algèbre Paul Dubreil et Marie-Paul Malliavin. Proceedings, 1987–1988. IV, 410 pages. 1989.

Vol. 1405: S. Dolecki (Ed.), Optimization. Proceedings, 1988. V, 223 pages. 1989. Vol. 1406: L. Jacobsen (Ed.), Analytic Theory of Continued Fractions III. Proceedings, 1988. VI, 142 pages. 1989.

Vol. 1407: W. Pohlers, Proof Theory. VI, 213 pages. 1989.

Vol. 1408: W. Lück, Transformation Groups and Algebraic K-Theory. XII, 443 pages. 1989.

Vol. 1409: E. Hairer, Ch. Lubich, M. Roche. The Numerical Solution of Differential-Algebraic Systems by Runge-Kutta Methods. VII, 139 pages. 1989.

Vol. 1410: F.J. Carreras, O. Gil-Medrano, A.M. Naveira (Eds.), Differential Geometry. Proceedings, 1988. V, 308 pages. 1989.

Vol. 1411: B. Jiang (Ed.), Topological Fixed Point Theory and Applications. Proceedings. 1988. VI, 203 pages. 1989.

Vol. 1412: V.V. Kalashnikov, V.M. Zolotarev (Eds.), Stability Problems for Stochastic Models. Proceedings, 1987. X, 380 pages. 1989.

Vol. 1413: S. Wright, Uniqueness of the Injective III_1 Factor. III, 108 pages. 1989.

Vol. 1414: E. Ramirez de Arellano (Ed.), Algebraic Geometry and Complex Analysis. Proceedings, 1987. VI, 180 pages. 1989.

Vol. 1415: M. Langevin, M. Waldschmidt (Eds.), Cinquante Ans de Polynômes. Fifty Years of Polynomials. Proceedings, 1988. IX, 235 pages. 1990.

Vol. 1416: C. Albert (Ed.), Géométrie Symplectique et Mécanique. Proceedings, 1988. V, 289 pages. 1990.

Vol. 1417: A.J. Sommese, A. Biancofiore, E.L. Livorni (Eds.), Algebraic Geometry. Proceedings, 1988. V, 320 pages. 1990.

Vol. 1418: M. Mimura (Ed.), Homotopy Theory and Related Topics. Proceedings, 1988. V, 241 pages. 1990.

Vol. 1419: P.S. Bullen, P.Y. Lee, J.L. Mawhin, P. Muldowney, W.F. Pfeffer (Eds.), New Integrals. Proceedings, 1988. V, 202 pages. 1990.

Vol. 1420: M. Galbiati, A. Tognoli (Eds.), Real Analytic Geometry. Proceedings, 1988. IV, 366 pages. 1990.

Vol. 1421: H.A. Biagioni, A Nonlinear Theory of Generalized Functions, XII, 214 pages. 1990.

Vol. 1422: V. Villani (Ed.), Complex Geometry and Analysis. Proceedings, 1988. V, 109 pages. 1990.

Vol. 1423: S.O. Kochman, Stable Homotopy Groups of Spheres: A Computer-Assisted Approach. VIII, 330 pages. 1990.

Vol. 1424: F.E. Burstall, J.H. Rawnsley, Twistor Theory for Riemannian Symmetric Spaces. III, 112 pages. 1990.

Vol. 1425: R.A. Piccinini (Ed.), Groups of Self-Equivalences and Related Topics. Proceedings, 1988. V, 214 pages. 1990.

Vol. 1426: J. Azéma, P.A. Meyer, M. Yor (Eds.), Séminaire de Probabilités XXIV, 1988/89. V, 490 pages. 1990.

Vol. 1427: A. Ancona, D. Geman, N. Ikeda, École d'Eté de Probabilités de Saint Flour XVIII, 1988. Ed.: P.L. Hennequin. VII, 330 pages. 1990.

Vol. 1428: K. Erdmann, Blocks of Tame Representation Type and Related Algebras. XV. 312 pages. 1990.

Vol. 1429: S. Homer, A. Nerode, R.A. Platek, G.E. Sacks, A. Scedrov, Logic and Computer Science. Seminar, 1988. Editor: P. Odifreddi. V, 162 pages. 1990.

Vol. 1430: W. Bruns, A. Simis (Eds.), Commutative Algebra. Proceedings. 1988. V, 160 pages. 1990.

Vol. 1431: J.G. Heywood, K. Masuda, R. Rautmann, V.A. Solonnikov (Eds.), The Navier-Stokes Equations – Theory and Numerical Methods. Proceedings, 1988. VII, 238 pages. 1990.

Vol. 1432: K. Ambos-Spies, G.H. Müller, G.E. Sacks (Eds.), Recursion Theory Week. Proceedings, 1989. VI, 393 pages. 1990.

Vol. 1433: S. Lang, W. Cherry, Topics in Nevanlinna Theory. II, 174 pages.1990.

Vol. 1434: K. Nagasaka, E. Fouvry (Eds.), Analytic Number Theory. Proceedings, 1988. VI, 218 pages. 1990.

Vol. 1435: St. Ruscheweyh, E.B. Saff, L.C. Salinas, R.S. Varga (Eds.), Computational Methods and Function Theory. Proceedings, 1989. VI, 211 pages. 1990.

Vol. 1436: S. Xambó-Descamps (Ed.), Enumerative Geometry. Proceedings, 1987. V, 303 pages. 1990.

Vol. 1437: H. Inassaridze (Ed.), K-theory and Homological Algebra. Seminar, 1987–88. V, 313 pages. 1990.

Vol. 1438: P.G. Lemarié (Ed.) Les Ondelettes en 1989. Seminar. IV, 212 pages. 1990.

Vol. 1439: E. Bujalance, J.J. Etayo, J.M. Gamboa, G. Gromadzki. Automorphism Groups of Compact Bordered Klein Surfaces: A Combinatorial Approach. XIII, 201 pages. 1990.

Vol. 1440: P. Latiolais (Ed.), Topology and Combinatorial Groups Theory. Seminar, 1985–1988. VI, 207 pages. 1990.

Vol. 1441: M. Coornaert, T. Delzant, A. Papadopoulos. Géométrie et théorie des groupes. X, 165 pages. 1990.

Vol. 1442: L. Accardi, M. von Waldenfels (Eds.), Quantum Probability and Applications V. Proceedings, 1988. VI, 413 pages. 1990.

Vol. 1443: K.H. Dovermann, R. Schultz, Equivariant Surgery Theories and Their Periodicity Properties. VI, 227 pages. 1990.

Vol. 1444: H. Korezlioglu, A.S. Ustunel (Eds.), Stochastic Analysis and Related Topics VI. Proceedings, 1988. V, 268 pages. 1990.

Vol. 1445: F. Schulz, Regularity Theory for Quasilinear Elliptic Systems and – Monge Ampère Equations in Two Dimensions. XV, 123 pages. 1990.

Vol. 1446: Methods of Nonconvex Analysis. Seminar, 1989. Editor: A. Cellina. V, 206 pages. 1990.

Vol. 1447: J.-G. Labesse, J. Schwermer (Eds), Cohomology of Arithmetic Groups and Automorphic Forms. Proceedings, 1989. V, 358 pages. 1990.

Vol. 1448: S.K. Jain, S.R. López-Permouth (Eds.), Non-Commutative Ring Theory. Proceedings, 1989. V, 166 pages. 1990.

Vol. 1449: W. Odyniec, G. Lewicki, Minimal Projections in Banach Spaces. VIII, 168 pages. 1990.

Vol. 1450: H. Fujita, T. Ikebe, S.T. Kuroda (Eds.), Functional-Analytic Methods for Partial Differential Equations. Proceedings, 1989. VII, 252 pages. 1990.

Vol. 1451: L. Alvarez-Gaumé, E. Arbarello, C. De Concini, N.J. Hitchin, Global Geometry and Mathematical Physics. Montecatini Terme 1988. Seminar. Editors: M. Francaviglia, F. Gherardelli. IX, 197 pages. 1990.

Vol. 1452: E. Hlawka, R.F. Tichy (Eds.), Number-Theoretic Analysis. Seminar, 1988–89. V, 220 pages. 1990.

Vol. 1453: Yu.G. Borisovich, Yu.E. Gliklikh (Eds.), Global Analysis – Studies and Applications IV. V, 320 pages. 1990.

Vol. 1454: F. Baldassari, S. Bosch, B. Dwork (Eds.), p-adic Analysis. Proceedings, 1989. V, 382 pages. 1990.

Vol. 1455: J.-P. Françoise, R. Roussarie (Eds.), Bifurcations of Planar Vector Fields. Proceedings, 1989. VI, 396 pages. 1990.

Vol. 1456: L.G. Kovács (Ed.), Groups – Canberra 1989. Proceedings. XII, 198 pages. 1990.

Vol. 1457: O. Axelsson, L.Yu. Kolotilina (Eds.), Preconditioned Conjugate Gradient Methods. Proceedings, 1989. V, 196 pages. 1990.

Vol. 1458: R. Schaaf, Global Solution Branches of Two Point Boundary Value Problems. XIX, 141 pages. 1990.

Vol. 1459: D. Tiba, Optimal Control of Nonsmooth Distributed Parameter Systems. VII, 159 pages. 1990.

Vol. 1460: G. Toscani, V. Boffi, S. Rionero (Eds.), Mathematical Aspects of Fluid Plasma Dynamics. Proceedings, 1988. V, 221 pages. 1991.

Vol. 1461: R. Gorenflo, S. Vessella, Abel Integral Equations. VII, 215 pages. 1991.

Vol. 1462: D. Mond, J. Montaldi (Eds.), Singularity Theory and its Applications. Warwick 1989, Part I. VIII, 405 pages. 1991.

Vol. 1463: R. Roberts, I. Stewart (Eds.), Singularity Theory and its Applications. Warwick 1989, Part II. VIII, 322 pages. 1991.

Vol. 1464: D. L. Burkholder, E. Pardoux, A. Sznitman, Ecole d'Eté de Probabilités de Saint- Flour XIX-1989. Editor: P. L. Hennequin. VI, 256 pages. 1991.

Vol. 1465: G. David, Wavelets and Singular Integrals on Curves and Surfaces. X, 107 pages. 1991.

Vol. 1466: W. Banaszczyk, Additive Subgroups of Topological Vector Spaces. VII, 178 pages. 1991.

Vol. 1467: W. M. Schmidt, Diophantine Approximations and Diophantine Equations. VIII, 217 pages. 1991.

Vol. 1468: J. Noguchi, T. Ohsawa (Eds.), Prospects in Complex Geometry. Proceedings, 1989. VII, 421 pages. 1991.

Vol. 1469: J. Lindenstrauss, V. D. Milman (Eds.), Geometric Aspects of Functional Analysis. Seminar 1989-90. XI, 191 pages. 1991.

Vol. 1470: E. Odell, H. Rosenthal (Eds.), Functional Analysis. Proceedings, 1987-89. VII, 199 pages. 1991.

Vol. 1471: A. A. Panchishkin, Non-Archimedean L-Functions of Siegel and Hilbert Modular Forms. VII, 157 pages. 1991.

Vol. 1472: T. T. Nielsen, Bose Algebras: The Complex and Real Wave Representations. V, 132 pages. 1991.

Vol. 1473: Y. Hino, S. Murakami, T. Naito, Functional Differential Equations with Infinite Delay. X, 317 pages. 1991.

Vol. 1474: S. Jackowski, B. Oliver, K. Pawałowski (Eds.), Algebraic Topology, Poznań 1989. Proceedings. VIII, 397 pages. 1991.

Vol. 1475: S. Busenberg, M. Martelli (Eds.), Delay Differential Equations and Dynamical Systems. Proceedings, 1990. VIII, 249 pages. 1991.

Vol. 1476: M. Bekkali, Topics in Set Theory. VII, 120 pages. 1991.

Vol. 1477: R. Jajte, Strong Limit Theorems in Noncommutative L_2-Spaces. X, 113 pages. 1991.

Vol. 1478: M.-P. Malliavin (Ed.), Topics in Invariant Theory. Seminar 1989-1990. VI, 272 pages. 1991.

Vol. 1479: S. Bloch, I. Dolgachev, W. Fulton (Eds.), Algebraic Geometry. Proceedings, 1989. VII, 300 pages. 1991.

Vol. 1480: F. Dumortier, R. Roussarie, J. Sotomayor, H. Żoładek, Bifurcations of Planar Vector Fields: Nilpotent Singularities and Abelian Integrals. VIII, 226 pages. 1991.

Vol. 1481: D. Ferus, U. Pinkall, U. Simon, B. Wegner (Eds.), Global Differential Geometry and Global Analysis. Proceedings, 1991. VIII, 283 pages. 1991.

Vol. 1482: J. Chabrowski, The Dirichlet Problem with L^2-Boundary Data for Elliptic Linear Equations. VI, 173 pages. 1991.

Vol. 1483: E. Reithmeier, Periodic Solutions of Nonlinear Dynamical Systems. Vi, 171 pages. 1991.

Vol. 1484: H. Delfs, Homology of Locally Semialgebraic Spaces. IX, 136 pages. 1991.

Vol. 1485: J. Azéma, P. A. Meyer, M. Yor (Eds.), Séminaire de Probabilités XXV. VIII, 440 pages. 1991.

Vol. 1487: E. Freitag, Singular Modular Forms and Theta Relations. VI, 172 pages. 1991.

Vol. 1490: K. Metsch, Linear Spaces with Few Lines. XIII, 196 pages. 1991.